0100

ELECTRICAL WIRING:
Principles and Practices

ELECTRICAL WIRING:
Principles and Practices

Clyde N. Herrick

San Jose City College
San Jose, California

PRENTICE-HALL, INC. ENGLEWOOD CLIFFS, NEW JERSEY

Library of Congress Cataloging in Publication Data
HERRICK, CLYDE N
    Electrical wiring.

    1. Electric wiring. I. Title
TK3201.H47   621.319'24   75-6634
ISBN 0-13-247676-2

© 1975 by Prentice-Hall, Inc. Englewood Cliffs, New Jersey

10   9   8   7   6   5

Printed in the United States of America

Prentice-Hall International, Inc., London
Prentice-Hall of Australia Pty, Ltd., Sydney
Prentice-Hall of Canada, Ltd., Toronto
Prentice-Hall of India Private Limited, New Delhi
Prentice-Hall of Japan, Inc., Tokyo

CONTENTS

PREFACE

This textbook presents a functional state-of-the-art teaching tool for classroom and shop instruction in electrical wiring. High-school and junior-college vocational students with a minimum background in electricity will find this text easy to understand and interesting to read. To facilitate learning and retention, profuse illustration, entertaining shop projects, and thought-provoking questions have been included. Unlike ordinary "wiring books" this text covers a broad spectrum, including basic electrical principles, devices, circuits, systems, wiring layouts, calculations, planning, estimating, and detailed installation procedures. In addition to residential wiring, the reader is also introduced to commercial wiring requirements and to the basic principles of automotive, marine, aeronautical, and telephone wiring systems.

It is assumed that the student has completed courses in arithmetic and high-school physics. A knowledge of elementary algebra will be helpful, although not essential to understanding the basic electrical laws and calculations that are explained in various chapters. Some knowledge of the use of hand tools such as augers, small saws, hand drills, hammers, chisels, cutters, and conduit (pipe) threaders will also be helpful. However, manual skills can be taught effectively in concurrent shop classes. Each student should be encouraged to obtain the latest edition of the National Electrical Code and to refer to those regulations frequently in both class and shop work. Students will also find it instructive to obtain copies of local electrical codes that may prevail.

This textbook is the outcome of extensive teaching experience, both on the part of the author and of his fellow instructors at San Jose City College, who have contributed numerous constructive criticisms and suggestions. In a significant sense, this book can be described as a team effort, although the individual members would choose to minimize the measure of their own contributions. The author is indebted to the sources credited in the text for their cooperation and provision of photographs, diagrams, and technical data. It is appropriate that this textbook be dedicated as a teaching tool to the instructors and students of our technical schools, high schools, and junior colleges. The author also wishes to express his appreciation to the two people who are most responsible for completion of the text: Margaret McAbee, the production editor, and Robert Mosher, the artist.

CLYDE N. HERRICK

ELECTRICAL WIRING:
Principles and Practices

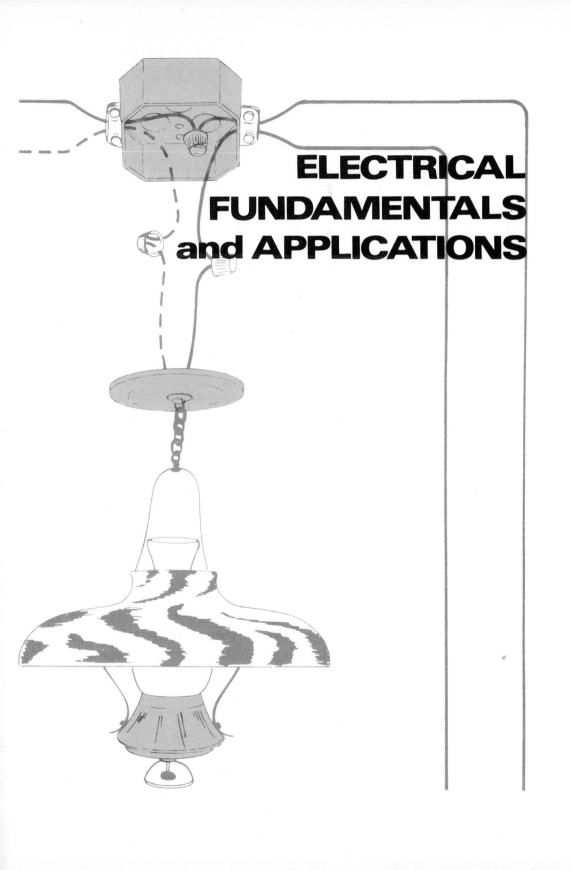

ELECTRICAL
FUNDAMENTALS
and APPLICATIONS

1 PRINCIPLES OF ELECTRICITY

1–1 BASIC PRINCIPLES

Electricity consists of a flow of charge units that are called *electrons*. Because these electrical charges are under pressure (in motion), they have energy. Copper wire is commonly used to carry these units called electrons from the source of electricity to the point of use.

When electrons flow in a copper wire, they are pushed along by a form of pressure that is called *voltage*. Voltage is an electrical force that is powerful enough to do various kinds of work. Electrical energy can light a room, for example, or heat a stove; do all kinds of work. For example, when electricity is supplied to a light bulb or incandescent lamp, its electrical energy changes into light energy. An electric motor changes or converts electricity into mechanical energy. A hot plate changes electricity into heat energy. By means of electrical energy a radio receiver converts electromagnetic waves into sound energy.

All of these devices—the light bulb, the electric motor, the hot plate, the radio receiver, must be connected to an electric power source by a suitable wiring system before they will work.

Figure 1-1 shows all of the lamps and appliances in a home. Almost all of the wiring system in a house is hidden inside of the walls—all concealed except for the cords that carry electricity from the wall outlets to appliances and lamps.

Unless the electrical wiring system for a house is well planned and installed, and then used correctly, various trouble symptoms can occur. For example, if an electrical wiring system is inadequate, or too skimpy to do the job, fuses may blow out or circuitbreakers may trip for no apparent reason.

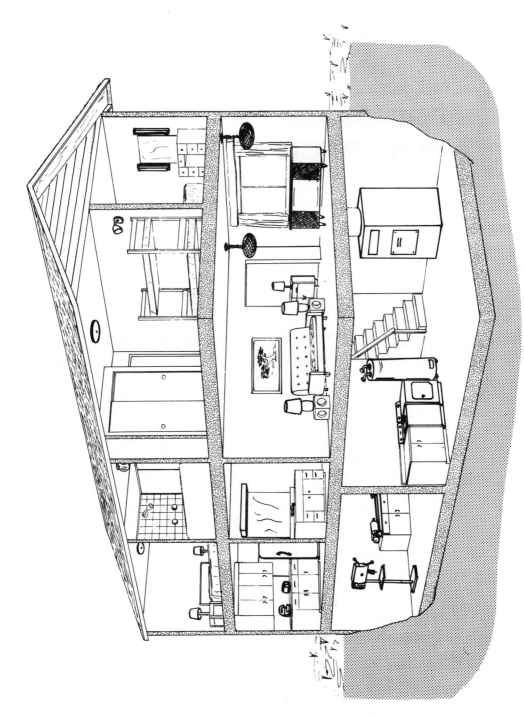

Figure 1-1 Cutaway view of lamps and appliances in a residence

Lights that burn brightly on one occasion may blink, become dim, or flicker when other appliances are turned on, and the appliances may fail to heat up normally. Television pictures may shrink or "lose sync." The air conditioner may slow down or even stall when several appliances are turned on at the same time. These problems can be avoided if the electrical wiring system is well planned and carefully installed by an electrician who uses good wiring practices.

1–2 SIX TERMS OF ELECTRICITY

The copper wires that carry electricity are usually grouped together into what is called an *electrical circuit*. An electrical circuit consists of all of the wires that conduct the flow of electrons from the source of electricity to the point where work is to be done by electrical devices.

Figure 1-2 shows a basic electric circuit in its ON (operating/closed) and OFF (not operating/open) conditions. This figure shows both a picture and, below that, a schematic diagram of the circuit. Electricians always work from schematic diagrams.

Note in Figure 1-2 that the lamp (the incandescent light bulb) does not glow unless there is a complete path for electrons to flow through the circuit. The electrons must be able to flow completely around the circuit inside the insulated copper wires.

A dry cell battery, such as that used in a large flashlight, is a source of direct current (DC). Its electrons flow in the same path and follow the same direction through the wiring circuit.

The voltage of a new carbon-zinc dry cell battery is approximately 1.5 volts. *Volts* is the short term that is used to state the amount of voltage in a circuit. It is indicated on schematic diagrams by the letter v or V. When the circuit in Figure 1-2 is completed or closed, by pushing the switch ON, this 1.5 volts causes electrical current to flow in the bulb or lamp. The switch acts as a bridge to close and complete the circuit. When the circuit in Figure 1-2 is open—that is when the switch is OFF—the voltage appears stalled at the ends of the open circuit.

The flow of electrons is rather like the flow of traffic—similar to cars driving up and down a road. The road has a bridge, and when the bridge is out of service for some reason, traffic must stop. When there is an "open" in the electrical circuit, the flow of electrons must stop.

Electrons are particles of negative electricity. Each electron is a tiny negative charge. The flow of electrons through the circuit is called *current*, as mentioned above. Current is measured in ampere units, or *amps*. The ampere

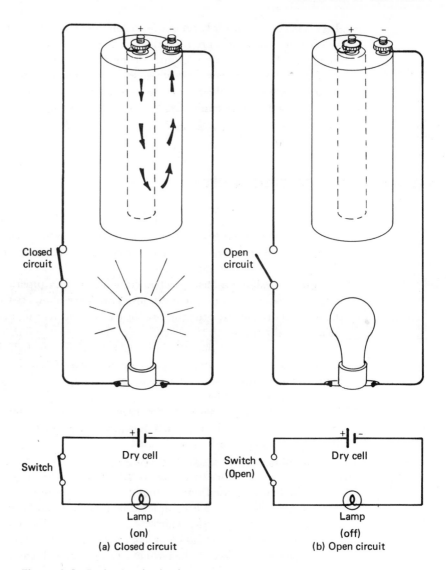

Figure 1-2   Basic electric circuit

is a term used to measure the strength of electric current. One ampere consists of 6.24×10^{18} (6,240,000,000,000,000,000) electrons flowing past one point of the wire during one second of time. It is indicated on schematic diagrams by the abbreviation amp, A, or AMP.

Note that there can be no current in a circuit unless voltage is applied from the electric source, because the flow of electrons is resisted and opposed by the copper wire or the lamp filament. Voltage must be applied from the electrical source in order to force the electrons through the *resistance* of a wire, filament, or other conductor of electricity. We measure resistance in ohm units, or *ohms*. This is indicated on electrical diagrams by the Greek letter omega (Ω). In the future, we will use just the words volts, amperes, and ohms, or their symbols, V, A, Ω—volts to measure the amount of force pushing the current, amperes to measure the amount of electricity coming through the circuit because of that voltage, and ohms to measure the amount of resistance put up by the wire and any devices connected to the circuit.

When electric current is forced through the resistance of the circuit and any appliance that is in use, physical work is performed. For example, the current from the dry-cell battery performs work because electrical energy is converted into both light and heat energy by the lamp filament. Note that the rate of doing work (changing energy) is called *power*. Power is measured in watt units, or *watts*. It is indicated by the letter w or W.

Power may be present in a circuit for a short time or for a long time. If power is present for a longer time, more work is done or more energy is converted. Electrical work is measured in *watt-hours*.

Every house, industrial, or institutional wiring system is connected to a watt-hour meter. A typical watt-hour meter that measures the amount of electrical work provided is shown in Figure 1-3. This watt-hour meter measures and displays on the five dials the amount of electrical energy that has been used over a period of time. Remember that work is equal to energy, and that power is equal to the amount of work that is done in one unit of time.

Figure 1-3 A watt-hour meter (Courtesy Westinghouse Electric Corp.)

1–3 TWO LAWS OF ELECTRICITY

Ohm's Law Voltage, current, and resistance are related by an important law of electricity called Ohm's law. Ohm's law states that the current in a circuit is directly proportional to the voltage applied, and inversely proportional to the resistance of the circuit. That is, the current will equal the voltage divided by the resistance. Ohm's law may be written

$$\text{Current} = \frac{\text{Voltage}}{\text{Resistance}}$$

or

$$\text{Amperes} = \frac{\text{Volts}}{\text{Ohms}}$$

or

$$I = \frac{E}{R}$$

which is simply an abbreviation of the idea that Input current (I) is equal to Electrical force (E) of the voltage divided by the Resistance (R). Using numbers, if the voltage is 220, and the resistance is 2, then the current will be 110. Using the dry cell battery shown in Figure 1-2 as our example again, if a battery supplies 1.5 V to the lamp filament, and the filament has a resistance of 6Ω, then the circuit is equal to .25 A or 1/4 ampere.

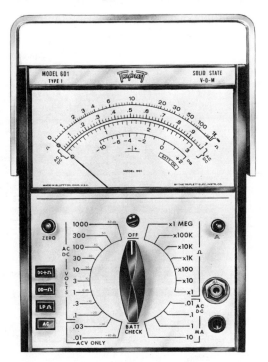

Figure 1-4 A volt ohm-milliammeter (Courtesy Triplett Electrical Instrument Co.)

Figure 1-4 shows a *volt-ohm-milliameter* (VOM) that measures volt, ampere, and ohm values. This VOM is a very useful instrument for the practical electrician.

Watt's (or the Power) Law Voltage and current are related to power by the second law of electricity, Watt's law. Watt's law shows that power (in watts) is equal to the product of voltage and current. That is, power equals volts times the amperes of current. Thus, we write

$$\text{Power} = \text{Voltage} \times \text{Current}$$

or

$$\text{Watts} = \text{Volts} \times \text{Amperes}$$

or

$$P = E \times I$$

The last is a shorthand way of writing Power equals the Electrical force of the voltage times current Input.

Again using numbers, if the current in a house is 110 A and the voltage is 220 V, we multiply these to find the watts—24,200.

If the dry cell shown in Figure 1-2 applies 1.5 volts to a lamp filament and the current in the circuit is 1/4 ampere (.25), then the power is equal to 0.375 watt. If the lamp stays lighted for 267 hours, about 100 watt-hours of electrical power will be converted into light and heat energy.

Since power is equal to current times the voltage (Watt's law), and current is equal to voltage divided by the resistance (Ohm's law), we can substitute the elements and state Watt's law in terms of current and resistance:

$$P = I^2 \times R$$

Similarly, we can state the power law in terms of voltage and resistance:

$$P = \frac{E^2}{R}$$

Figure 1-5 Summary of the basic electrical formulas

Again current, voltage, and resistance can be expressed in three ways:

$$I = \frac{E}{R} = \frac{P}{E} = \sqrt{\frac{P}{R}}$$

$$E = IR = \frac{P}{I} = \sqrt{PR}$$

$$R = \frac{E}{I} = \frac{P}{I^2} = \frac{E^2}{P}$$

The formula that we use in a particular circuit calculation depends on the values that we happen to have available. The diagram in Figure 1-5 will be helpful in case of doubt concerning a basic electrical relation. Each quarter circle contains the terms explained above.

1–4 BASIC ELECTRICAL MEASUREMENT UNITS

The practical system of electrical units includes the ampere, the ohm, the volt, and the watt. You will recall that the ampere is the measuring unit of current flow. It consists of 6.24×10^{18} electrons flowing every second past a given point in a circuit. The ohm is the measuring unit of resistance. It is the measure of a resistor's ability to oppose the flow of current. The volt is the unit of electromotive force (electrical pressure). It is defined as the potential difference (voltage) that causes 1 ampere to flow through a resistance of 1 ohm. The watt is the unit of power. It is defined as the product of 1 volt multiplied by 1 ampere.

In addition to these four units of measurement, the practical system of electrical units also includes the *Hertz*, the *henry* and the *farad*. The Hertz is the unit of frequency. It is indicated in a schematic diagram by the letters Hz. Frequency is the number of alternations per second of an alternating electric current. When alternating current (AC) goes through a positive value and then through a sequence or cycle of negative value during 1 second of time, its frequency is defined as 1 Hertz.

The *farad* (indicated by F) is the unit of electrical storage or capacitance. Capacitance is the property of a nonconductor that permits electrical storage when input and output are unequal. A farad is the amount of capacitance required to store 1 ampere-second when 1 volt is applied across the capacitor. (See Figure 1-6.)

The *henry* is the unit of inductance. Inductance is a property of an electrical circuit by which electromotive force is brought about by varying the current (pulsing) or because of the transfer effects of a nearby electric circuit. The henry is defined as the amount of inductance that permits current to increase at a rate of 1 ampere per second when 1 volt is applied across the inductor. (See Figure 1-7.)

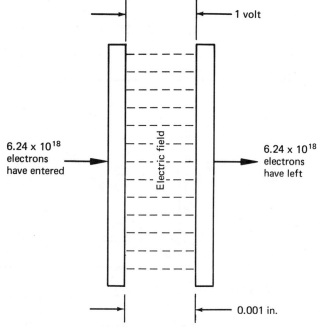

6.24 x 10^{18}
electrons
have entered

1 volt

Electric field

6.24 x 10^{18}
electrons
have left

0.001 in.

Area of each plate is 4.46 x 10^9 square inch
capacitance is 1 farad

(a)

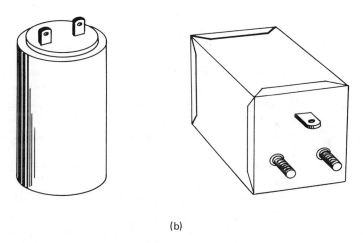

(b)

Figure 1-6 Electrical capacitors (a) principles of a 1-farad capacitor (b) typical motor-starting capacitors

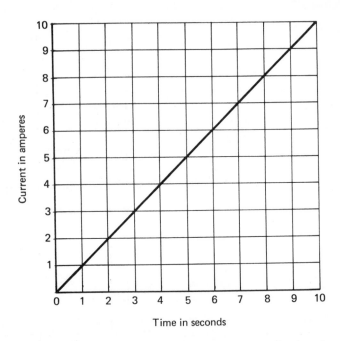

Figure 1-7 Rise of current in an ideal 1-Henry coil, when 1
volt is applied

Note also that the watt-second is a unit of electrical energy. It is equal to
the expenditure of 1 watt of power for 1 second. Electricians commonly use
the watt-hour unit; a watt-hour is equal to the expenditure of 1 watt of
power for 1 hour.

MULTIPLE AND SUBMULTIPLE UNITS

Whole units such as the volt, ampere, ohm, watt, Hertz, henry, farad, and
watt-hour are convenient units of measurement to use in most situations.
However, it is sometimes helpful to use multiple or submultiple measuring
units. Practical electricians may state the power demand of a central air-
conditioner, for example, either as 5,000 watts or as 5 kilowatts (5 kW).
A kilo is a multiple unit that denotes multiplication by 1,000. It is indicated
by *k* or K. Similarly, a meter such as the one shown in Figure 1-3 is generally
called a kilowatt-hour (kwh) meter.

Note that the volt-ohm-milliameter (VOM) shown in Figure 1-4 has
a milliampere range. Milli is a submultiple that denotes a thousandth part
of a unit or 0.001. That is, milli denotes division by 1,000. If a small night-
light, for example, draws 0.25 ampere, its current value may be stated as 250
milliamperes (250 mA).

A neon sign may operate at a potential of 10,000 volts. This value is also stated as 10 kilovolts (10 kV). However, a neon tube draws a very small current, perhaps 0.03 ampere (30 mA).

Small electrical storage devices called capacitors, which will be discussed in detail later, are used in various electrical applications. They have capacitance values on the order of thousandths of a farad, so the submultiple unit called the *micro*farad is commonly discussed. Micro means one-millionth of a part of some whole unit. A capacitor that is used to start a motor might have a value of 1,000 microfarads, or 1,000 μF, as this is shown on schematics.

Finally, note that *giga* (G) denotes 1 billion, as used in radar electrical systems.

1–5 HOW TO READ BASIC RECORDING INSTRUMENTS

Recording wattmeters (watt-hour meters) or kilowatt-hour meters have dials as shown in Figure 1-8. Note that the first dial reads counterclockwise, but the second dial reads clockwise, then the third dial reads counterclockwise again, and so on. The meter readings shown in Figure 1-8 are self-explanatory. These can be used for practice and as practical examples of meter reading. It is also helpful and informative to observe the kwh meter of the building that you are in, and to read that meter indication.

Electricians who are employed by electric power companies make use of *strip recorders*, as shown in Figure 1-9. The strip recorder is an instrument that displays either voltage or current and makes a record of variations over an extended period of time. A strip recorder is commonly used to monitor incoming line voltage at a factory during a 24-hour interval. If excessive voltage variation occurs, the power company might decide to install a larger line transformer or to run additional lines into the factory. Strip recordings are made by a pen-like stylus on a slowly moving strip of ruled paper. The rules on the paper may correspond to voltage values (such as 5 volts per line). The paper strip can be moved at rates up to 1 foot per hour.

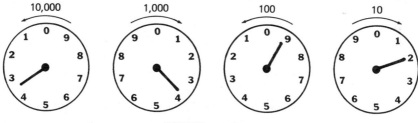

3392 Kilowatt-hours

Figure 1-8 Examples of dials for kilowatt-hour meters

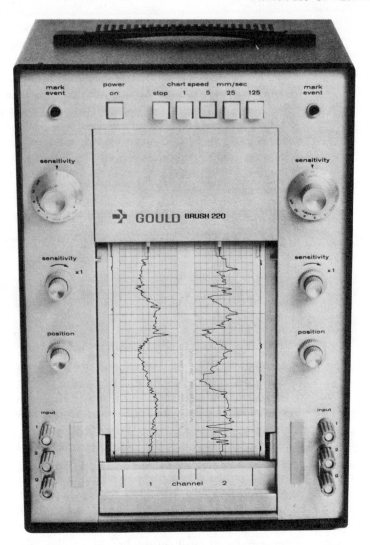

Figure 1-9 Two-channel general purpose recorder (Courtesy Gould Brush)

Note that a strip recorder provides much different information than a kwh meter. The strip recorder provides a record of voltage or current variations over time and does not indicate the amount of energy used. On the other hand, a kwh meter indicates the total energy used over a set period without giving information about voltage or current during that time. Electrical machinery operates inefficiently unless the supply voltage is maintained near its rated value. Overvoltage is just as undesirable as undervoltage. Electric light bulbs burn out quickly when they are operated above their rated voltage. The electric power company must maintain line voltage as nearly constant as possible.

1–6 BASIC ELECTRIC CIRCUITS

A *series circuit*, shown in Figure 1-10, has its component parts connected in a chain. Note that the current is the same at any point in the series circuit. The total resistance of the series circuit is equal to the sum of all of its component resistances. This can be written

$$R_{\text{total}} = R_1 + R_2 + R_3 + R_4 + R_5 + R_6 + R_7 + R_8$$

If any one lamp burns out in a series circuit, all lamps become dark, because the circuit becomes open. When the burned out lamp is replaced, all of the lamps will glow. It is possible to see a burned; out filament if the light bulb is transparent. The practical troubleshooting procedure is to replace the lamps one by one until the series string starts to glow. If you have a VOM or ohmmeter available, a filament can be checked for continuity. That is, a burned-out filament indicates infinite resistance on an ohmmeter scale, whereas a good filament will measure a definite number of ohms, such

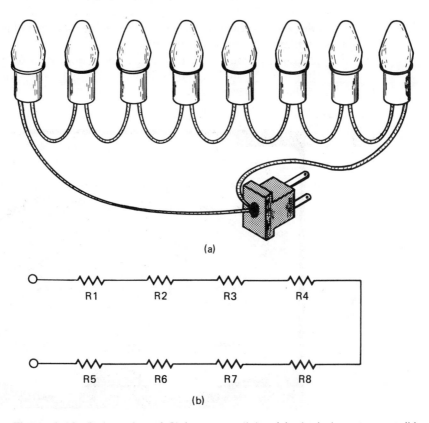

(a)

(b)

Figure 1-10 Series string of Christmas-tree lights (a) physical arrangement (b) electrical circuit

as 3 ohms. Figure 1-11 shows a scale for a VOM. Note that infinity is indicated at the left end of the ohms scale.

It is helpful to observe the settings used on a typical volt-ohm-milliammeter (VOM) for measurement of DC voltage values. Figure 1-12 shows how the function switch is set to +DC, and the range switch is set to 2.5 V in this example. The 2.5-V range is used when the voltage to be measured does not exceed 2.5 volts. For example, the voltage of an ordinary dry cell would be checked on this range. When the function switch is set to +DC, the pointer will deflect or move upscale (normal deflection) if the common (−) terminal is connected to the negative terminal of the dry cell, and the + terminal is connected to the positive terminal of the dry cell. However, if the dry cell is connected to the VOM terminals in opposite polarity, the pointer will deflect down-scale (below zero) unless the function switch is set to −DC. If down-scale deflection occurs, the pointer movement can be corrected either by reversing the terminal connections to the VOM, or by changing the function switch to opposite polarity setting.

Suppose next that a 6-volt storage battery is to be checked. If the VOM is operated on its 2.5-volt range, the pointer will overshoot the right-hand end of the scale, and the meter is very likely to be damaged. Therefore, the range switch must be set to its 10-volt position.

Again, suppose that a 12-volt storage battery is to be checked. Since 12 volts cannot be indicated on a 10-volt scale, the range switch must be set to its 50-volt position in this example. *The general rule is that the range to be used on the VOM must be equal to or greater than the voltage that is to be measured.* In

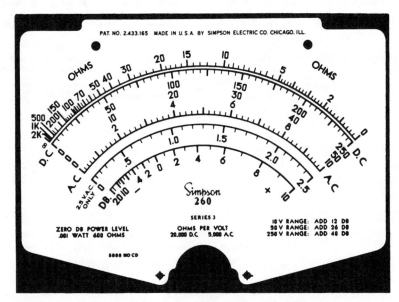

Figure 1-11   Scale plate for a VOM (Courtesy Simpson Electric Co.)

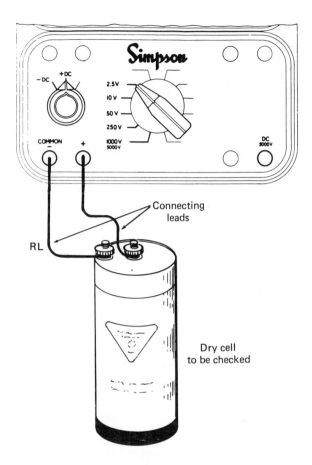

Figure 1-12   VOM control settings for operation on the 2.5
volt DC range (Courtesy Simpson Electric Co.)

order to protect the VOM when the voltage to be measured is not definitely
known, electricians first set the range switch to a very high value, such as
1,000 volts. Then, the connecting leads are applied to the voltage source,
and the setting of the range switch is reduced in steps until the VOM pointer
indicates above half-scale.

IN-SERIES CIRCUITS

Next, let us consider the basic principles of series circuits, as shown in
Figure 1-13. This consists of two dry cells connected in series-aiding sequence
(positive terminal to negative terminal). Series-aiding means that the voltage
of one cell adds to the voltage of the other cell. Since each dry cell provides
1.5 volts, the battery provides 3 volts. Two lamps are connected in series with
the battery, and electron flow is shown by the arrows. The schematic diagram

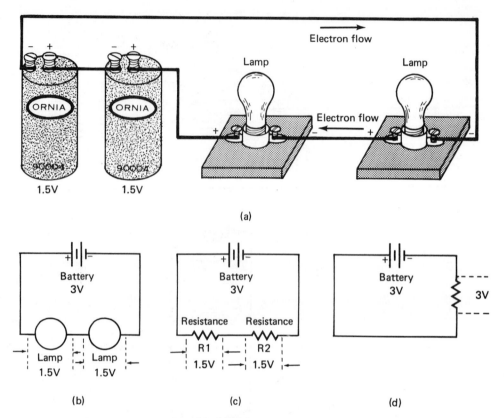

Figure 1-13 Basic principles of series circuits (a) picture diagram of circuit (b) schematic diagram, (c) first equivalent circuit (d) second equivalent circuit

in (b) corresponds to the picture diagram in (a). Since the lamp filament is a resistive element, an equivalent circuit can be drawn, as shown in (c). Finally, since the total load resistance is equal to $R_1 + R_2$, a second equivalent circuit can be drawn, as shown in (d).

Observe that the same type of lamp is used in both sockets in Figure 1-12. In turn, R_1 is equal to R_2 in (c). Since the same current value flows through R_1 and R_2, the voltage across R_1 is the same as the voltage across R_2. Furthermore, the sum of the voltage drops across R_1 and R_2 must be equal to the battery voltage. In other words, since the battery provides 3 volts, the voltage drop across each lamp is 1.5 volts. This shows that Ohm's law applies to each component in a circuit, just as it applies to the complete circuit.

With reference to Figure 1-13(c), let us suppose that the load $(R_1 + R_2)$ draws 333 mA from the 3-volt battery. In turn, the load draws 1 watt of power, and each lamp evidently draws 1/2 watt of power. Since the voltage drop across each lamp is 1.5 volts and the current value is 333 mA, it follows

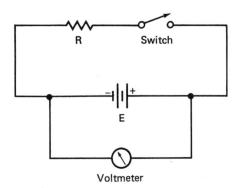

Figure 1-14 Voltmeter connections to measure battery voltage

that the resistance of the lamp filament is 4.5 ohms, approximately. However, as an experienced electrician would point out, this 4.5 ohms represents the "hot" resistance of the filament. A tungsten filament has a much lower value of "cold" resistance. As an illustration, if a filament has a "hot" resistance of 4.5 ohms, its "cold" resistance might be about 0.5 ohm. Ohmmeters measure the "cold" resistance of a filament, *excepting some low-voltage, low-power filaments that glow when tested due to the test current from the ohmmeter.*

It is very important for an electrician to understand the correct application of meters in electric circuits. The ohmmeter must be used only in a "dead" circuit or to test a component that has been removed from the circuit. In other words, the ohmmeter of the VOM must be connected across circuit points, or across a component, with zero volts present between the connection points.

The correct way to measure a source voltage with a voltmeter of a VOM is shown in Figure 1-14. Note here that the voltmeter is connected across the battery. If we wish to measure the voltage drop across resistor R, the voltmeter would be connected across the resistor and the switch would be closed (turned on). No voltmeter indication can be obtained across the resistor if the switch is open (off).

Next, the correct way of measuring the current in a circuit is shown in

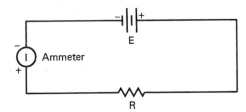

Figure 1-15 Ammeter connections to measure circuit current

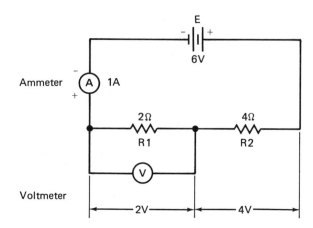

Figure 1-16   Correct connections of a voltmeter and an
ammeter in a series circuit

Figure 1-15. Note that the ammeter of the VOM is connected in series with
the circuit. It is essential to remember that *the ammeter must be connected only
in series with a circuit.*

Next, observe the connections of the ammeter and voltmeter in Figure
1-16. We see that the ammeter measures the current in the circuit; because
the current is the same at any point in a series circuit, it makes no difference
whether we connect the ammeter at one point in a circuit or at another
point. We see that the voltmeter measures the voltage across R_1. If the voltmeter were connected across R_2, it would indicate twice as much voltage.

Finally, let us carefully observe the DON'TS that we must avoid in using
meters:

1.   DON'T ever connect the ammeter across a voltage source such as E in Figure
     1-16. YOUR AMMETER WOULD BE DESTROYED INSTANTLY.

2.   DON'T connect the voltmeter in series with a circuit. If the voltmeter were con-
     nected in place of the ammeter in Figure 1-16, the voltmeter would simply
     indicate the battery voltage. If you doubled the values of the resistors, the
     voltmeter would still indicate the battery voltage.

3.   DON'T ever connect the ohmmeter across any points where voltage is present.
     For example, if you connect the ohmmeter across R_2 in Figure 1-16, YOUR
     OHMMETER WOULD BE DESTROYED INSTANTLY.

PARALLEL CIRCUITS

A *parallel circuit*, as given in Figure 1-17, is another important type of
electrical circuitry. Observe that each lamp is connected across the line.
Therefore, if one lamp burns out, the other lamps will continue to glow.

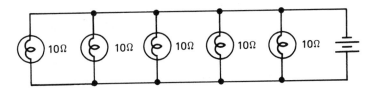

Figure 1-17 Example of a parallel circuit

This arrangement is used in most residential wiring systems. Observe that each lamp in Figure 1-17 has a filament resistance of 10 ohms. In turn, each lamp draws the same amount of current from the line. Since there are five lamps in the circuit, the total load resistance will be 2Ω. Of course, if the lamp filaments had various resistance values, the total load resistance could be calculated by using Ohm's law, that is, voltage divided by resistance equals current.

Figure 1-18 shows that Ohm's law applies to each branch of the circuit, as well as to the complete circuit. Note that the battery voltage is applied at a and b, and that this same voltage is applied across R_1, R_2, and R_3. The reading on ammeter At is equal to the sum of the readings on ammeters A1, A2, and A3. Each resistor wastes or dissipates power in accordance with its resistance value. That is, R1 dissipates EI_1 watts, R2 wastes EI_2 watts, and R3 dissipates EI_3 watts. In turn, the battery must supply power in an amount that is equal to the sum of three resistance or dissipation values.

If we keep the basic principles of series and parallel circuit operation in mind as we continue our study of electrical wiring, we will be able to analyze system circuitry without much difficulty.

An electrician must be able to use many hand and power tools in the course of doing his job. Some tools are basic to all electrical areas, while others are specific to certain areas. The tools shown in Fig. 1-19 are those most com-

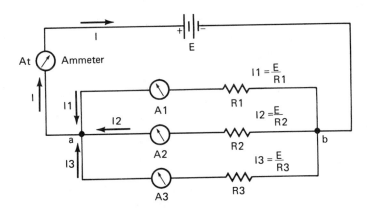

Figure 1-18 Current relations in a parallel circuit

monly used by the electrician. We will briefly describe the tools here and will refer to this page as tools are discussed throughout the text.

The *claw hammer* is used to drive nails.

The *brace and bit* is used to manually drill holes in wood.

The *key-hole saw* is used to enlarge a hole in wood.

The *neon test light* is used to determine if there is a voltage on a line.

The *folding rule* is used to measure distance.

The *chisel* is used to remove wood from a surface, or to enlarge a hole in wood.

The *lever-jaw wrench* is used as pliers, lock wrench or as pipe wrench.

Linesmen's pliers are used to cut wire and to tighten or hold a nut.

Wire strippers are used to strip the insulation from wire.

Multi-purpose tool is used to cut and strip wire, to attach terminals and lugs.

The *hack-saw* is used to cut metal objects such as conduit.

The *linesmen's voltmeter* is used to measure the value of the voltage on a line.

The *fuse puller* is a plastic or phenolic-insulated tool that is used to remove cartridge-type fuses.

The *jack-knife* is used by an electrician for trimming or cutting insulation from wire.

The *screwdriver* is used to tighten screws or to hold a screw on while a nut is being tightened.

A *fish-tape* is used for fishing wires through a wall or floor space and through conduit.

The *pipe cutter* is used to cut conduit.

The *conduit bending tool* (sometimes called a *hickey*), is used by the electrician to put bends in conduit.

STUDENT EXERCISE 1-A

The steps in this exercise provide practical experience in voltmeter application with basic series circuits.

1. Obtain one 12-volt lantern battery, eight miniature sockets (flashlight screw-base type), eight 1-1/2 volt flashlight bulbs, eight 3-volt flashlight bulbs, off-on switch, small roll of bell wire, and a DC voltmeter with ranges up to 15 volts or more.

2. Using the bell wire, connect the battery, switch, and sockets in a series circuit. (See Figure 1-10 for details.)

3. Insert the eight 1-1/2 volt bulbs into the sockets. Note that all bulbs light with equal brightness when the switch is closed (on), and that all bulbs are dark when the switch is open (off).

Hammer	Drill bit	Brace	Keyhole saw	Test light
Folding rule	Chisel	Lever-jaw wrench	Linesman's pliers	Wire stripper
Multi-purpose tool	Hack saw		Voltmeter	Fuse puller
Jack knife	Screwdriver	Fish tape and reel	Pipe cutter	Conduit bender

Figure 1-19 A basic set of electrician's tools

4. With the switch open, measure the voltage across the switch terminals. Use the 15-volt range of the voltmeter in this test.

5. With the switch closed, measure the voltage across the switch terminals.

6. With the switch closed, measure the voltage across each socket in turn. Use a 2- or 3-volt range in this test.

7. Replace the 1-1/2 volt bulbs with 3-volt bulbs and note the resulting drop in brightness.

8. With the switch closed, measure the voltage across each socket in turn.

9. Using the Ohm's law formula, try to calculate the result of replacing one of the 3-volt bulbs with a 1-1/2 volt bulb.

10. Insert a 1-1/2 volt bulb in place of a 3-volt bulb to check your calculations.

STUDENT EXERCISE 1-B

The steps in this exercise provide experience in voltmeter application with parallel circuits.

1. Obtain two No. 6 dry cells, eight miniature sockets (flashlight screw-base type), eight 4-1/2 volt flashlight bulbs, eight 3-volt flashlight bulbs, off-on switch, small roll of bell wire, and a DC voltmeter with ranges up to 5 volts or more.

2. Connect the sockets in parallel. (See Figure 1-17.)

3. Connect the dry cells to form a 3-volt battery (positive terminal of one cell to negative terminal of the other cell). Check the battery voltage with the voltmeter.

4. Connect the battery to the parallel lamp circuit, inserting the switch into one of the wires or leads from the battery to the lamps.

5. Insert the eight 3-volt bulbs into the sockets. Note that all bulbs light with equal brightness when the switch is closed, and that all bulbs are dark when the switch is open.

6. With the switch open, measure the voltage across the switch terminals. Use the 5-volt range of the voltmeter in this test.

7. With the switch closed, measure the voltage across the switch terminals.

8. With the switch closed, measure the voltage across each socket in turn.

9. Replace the 3-volt bulbs with 4-1/2 volt bulbs and note the resulting drop in brightness.

10. With the switch closed, measure the voltage across each socket in turn.

11. Using Watt's (power) law formula, try to calculate the result of replacing one of the 4-1/2 volt bulbs with a 3-volt bulb.

12. Insert a 3-volt bulb in place of a 4-1/2 volt bulb to check your calculations.

QUESTIONS AND EXERCISES

1. Define resistance as it applies to electricity.

2. What is electric current?

3. What is the unit of power in an electrical circuit?

4. What is the function of a watt-hour meter?

5. How much current will flow through the ohmmeter resistance of an electric heater that is connected to a 115-volt outlet?

6. How much current will flow through a 100-watt lamp connected to a 115-volt supply?

7. What is the unit of current?

8. What is the unit of energy?

9. What is the unit of resistance?

10. If one lamp in a series circuit burns out, why do all the lamps become dark?

11. How is an ohmmeter used to measure resistance in a circuit?

12. How is a voltmeter connected to measure a voltage drop?

13. How is an ammeter connected to measure current?

14. What are the characteristics of a parallel circuit?

15. What letters do we use to indicate: (a) current, (b) voltage, (c) resistance, and (d) power?

16. With reference to Fig. 1-1, how many lights and appliances are provided in the house?

17. What are the readings of the watt-hour and kilowatt-hour meters shown in Fig. 1-20?

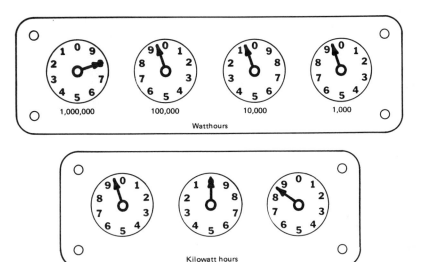

Figure 1-20

2 DC AND AC OPERATION

2–1 DC AND AC RELATIONSHIPS

Direct current (DC) is characterized by electron flow in one direction only, as depicted in Figure 2–1. On the other hand, alternating current (AC) is characterized by alternate flow in two directions, as shown in Figure 2–2. We will suppose that the switch is left in position 1 for a certain length of time, such as 1 second. Then, the switch is thrown to position 2 and left in this position for 1 second. This procedure is continued indefinitely. It is evident that electrons from the battery flow through the resistor (R) in the left-hand direction for the next second. This is the simplest form of AC current. As shown in Figure 2–3, the power dissipated by resistor R is the same as if the switch were left in one of its positions indefinitely. That is, the battery voltage is always applied to the resistor, although its polarity reverses at intervals.

2–2 FREQUENCY

Alternating current has a frequency that is defined as the number of times per second that the current goes through a complete sequence or cycle. With reference to Figure 2–3, the current goes through a positive excursion in 1 second, then goes through a negative excursion in 1 second. Thus, the current goes through a complete cycle in 2 seconds. In turn, the frequency of this alternating current is 1/2 Hertz (Hz). Note that Hertz is defined as cycles per second. A complete cycle consists of a positive half-cycle and a negative half-cycle.

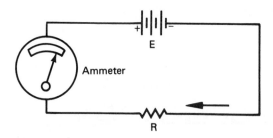

Figure 2-1 Electrons flow in one direction
only in a DC circuit

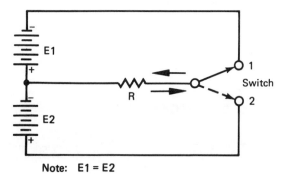

Note: E1 = E2

Figure 2-2   An AC demonstration circuit

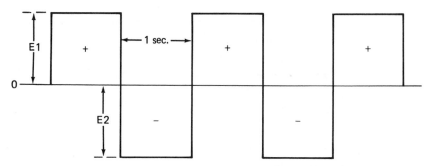

Figure 2-3   Graph of current in the circuit of Fig. 2-2

Power systems in the United States all operate at 60 Hz, although power
sources in airplane and certain specialized installations may operate at
higher frequencies.

Fluorescent lighting circuits sometimes operate at 400 Hz and some
industrial installations operate at 180 Hz. Military equipment is operated
at 400 Hz in various cases. As explained in detail later, the chief advantage
of AC over DC current is that alternating current can be transformed to
higher or lower voltages easily. Moreover, one advantage of higher power

frequencies is the decrease in size and weight of the transformers and power-conversion devices.

Just as resistance values may be expressed in multiple units, so may frequency values. For example, a high-fidelity amplifier ordinarily operates at up to 20,000 Hz (20 kHz). A radio broadcast receiver might operate at a frequency of 1,000,000 Hz (1 MHz); mega (M) denotes multiplication by 1 million. Again, a radar transmitter could operate at a frequency of 1,000,000,000 Hz (1 GHz); giga (G) as you recall, denotes multiplication by 1 billion. As practical electricians, we will seldom be called upon to work with circuits that operate at a frequency other than 60 Hz. Of course, it is important for us to know that fluorescent lighting and other installations may operate at higher frequencies.

2–3 SINUSOIDAL ALTERNATING CURRENT

The graph of alternating current shown in Figure 2–3 is called a square wave because of its shape. Square waveforms are used only in special applications. Public utilities supply alternating current with a *sinusoidal waveform*, as shown in Figure 2–4. A sinusoid is also called a sine wave. In your physics courses, a sine wave was described as simple harmonic motion. Figure 2–5 shows that a sine wave has a maximum (peak) value on each half-cycle. In other words, its positive peak value is equal to its negative peak value. At any particular instant, the AC waveform can be regarded as a DC waveform. In particular, a sine wave has an *effective* value at 70.7% of peak, as indicated in Figure 2–5. Consider carefully the significance of this effective value.

The effective value of a sine wave is the equivalent voltage of a DC source that will provide the same amount of light or power as the AC source. For example, a conventional residential wiring system operates at 117 volts effective value. This means that a lamp will burn at the same brightness when connected to the 117-volt AC line as if it were connected to a 117-volt DC source. Again, an electric heater will produce the same amount of heat, whether

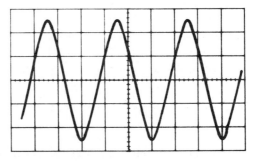

Figure 2-4 Photo of a 60-Hz sine wave from oscilloscope screen

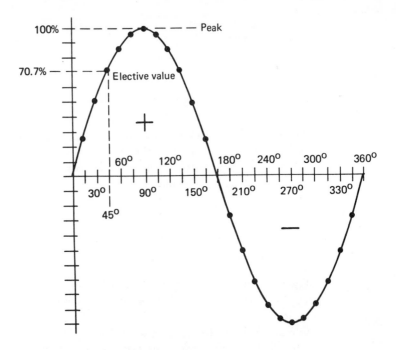

Figure 2-5   Detail of a sine waveform

it is connected to the 117-volt AC line or to a 117-volt DC source. The effective value of a sine wave is also called its *root-mean-square* (rms) *value.* Note that a VOM (Figure 1–4) provides both DC-voltage and rms-voltage AC scales. AC voltage cannot be measured on the DC function of a VOM.

Again, referring to Figure 2–5, observe that the peak value (100%) of the sine wave is greater than its effective, or rms, value. Since the rms value is equal to 70.7% of the peak value, it follows that the peak value is equal to 1.414 times the rms value. For instance, a 117-volt rms line has a peak voltage of about 165 volts. Note also in Figure 2–5 that a complete cycle of an alternating current contains 360°. In turn, a half-cycle contains 180°. The effective value of the sine wave occurs at 45° (or at 225°). Note that the effective value is a point, or height, on the sine wave. These facts will be of basic importance when we consider the amount of power drawn by fluorescent lamps, motors, and other electrical devices.

2–4 PHASE RELATIONS IN AC CIRCUITS

When sine wave voltage is applied to a lamp, or other load, a sine wave current flows through the load. In the case of a resistive load, the voltage and current waveforms are in phase with each other, as shown in Figure 2–6. The term "in phase" means that the current (I) goes through zero and goes through its peak value (1,000 W) at the same instants that the voltage

(E) goes through zero and goes through its peak value. In the left part of this graph, the power is indicated by the top curved line (P), the voltage is indicated by the middle curved line (E), and the current is indicated by the bottom curved line I. The instantaneous power flowing in this circuit ranges from zero at the baseline of the graph up to 1000 W at peak power. At right, the true power is shown as being equal to rms voltage multiplied by rms current—or 500 W in this example. Observe that this is the average real power in the load. On the other hand, instantaneous power (P) is obtained by multiplying instantaneous voltage values by corresponding instantaneous current values, varies from zero to 1,000 watts. This power waveform has a sine wave variation, or outline, enclosing the shaded area.

Note also in Figure 2–6 that all the power shown in the diagram is positive. Positive power is called *true power*, *real power*, or *in-phase power*. True power is shown above the zero axis. The significance of real power is that it converts electrical energy into some other form of energy. (Later we will learn of another type of power that cannot convert electrical energy into any other form of energy.) Observe in Figure 2–6 that the power waveform has twice the frequency of the voltage waveform or the current waveform. If the voltage waveform has a frequency of 60 Hz, the power waveform will have a frequency of 120 Hz. (Recall that power values are related only to voltage and current values—power is independent of frequency.)

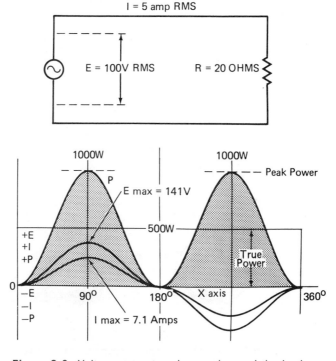

Figure 2-6 Voltage, current, and power in a resistive load

2–5 REACTIVE POWER

In Figure 2–7(a), a light bulb is connected in series with a capacitor to a DC-voltage source and in (b) to an AC-voltage source. A capacitor, such as a motor-starting capacitor, can be employed in this example. When connected to a DC source, the lamp does not glow because the capacitor blocks the direct-current flow. The lamp does glow when connected to the AC source because the capacitor permits AC current to flow in the circuit. The capacitor is first charged positively, and then negatively, by the half-cycles of alternating current. This reversing charge of current flows back and forth in the line and energizes the lamp filament.

Figure 2–8 shows the phase relations in a basic capacitive circuit. The capacitor is connected directly to an AC-voltage source, as shown by the symbol $\sim$. Instruments such as an oscilloscope can show that the current and voltage waveforms are 90° out of phase, as in the graph. That is, a capacitor has the property of shifting the current 90° with respect to the phase of the applied voltage. Note too that the current goes through its peak value before the voltage goes through its peak value. For this reason, electricians say that the current *leads* the voltage in a capacitive circuit. Note carefully that the power is half positive and half negative in Figure 2–8. Therefore, the positive/negative powers cancel each other over the

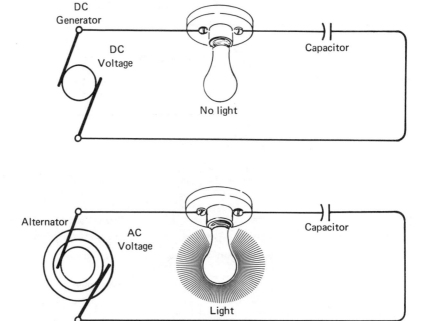

Figure 2-7   A series capacitor permits alternating current flow in a circuit

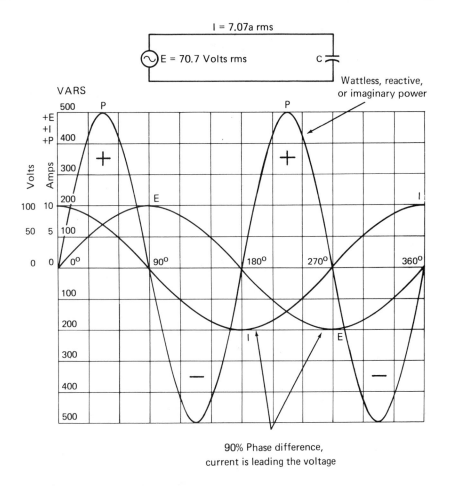

Figure 2-8 Phase relations in a basic capacitive circuit

complete cycle, and average power in the circuit is zero. This kind of power is called wattless, or *imaginary power*. It does no work. This is the result of the 90° phase difference between the voltage and current in the circuit. Wattless power merely surges back and forth in a circuit.

Although the average power is zero in the circuit of Figure 2–8, a certain value of wattless or *reactive power* is present. It is VARS and can be measured with a VAR meter. VAR means volt-amperes-reactive. One VAR, that is, one unit of reactive power, is numerically equal to 1 watt, but the distinction is that the VAR cannot do work. In the graph of Fig. 2–8, the reactive power is equal to 500 VARS. Next, returning to the AC circuit shown in Figure 2–7 (b), we have a series circuit that consists of the line and lamp, which acts as a resistor, and a capacitor connected to an AC source. In Figure 2–9, the phase relations of voltage, current, and power are shown for this

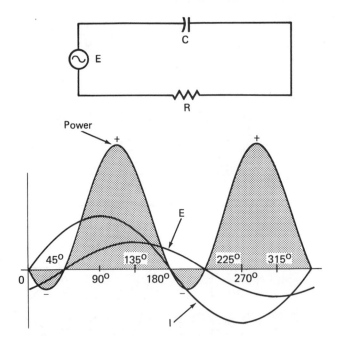

Figure 2-9   Phase relations in an RC circuit

Figure 2-10   An electrician's volt-amp-wattmeter (Courtesy Simpson Electric Company)

resistance-capacitance configuration. Consider the circuit action that is taking place.

Note that there is more positive power than negative power shown in the graph for the circuit. This is indicated by the difference between the large tinted area above the baseline (positive) and the shallow cup-like tinted area below the horizontal baseline (negative). For this reason there is both real power and reactive power present. Electricians may use an AC volt-amp-wattmeter, as in Figure 2–10, to measure the voltage, current and real power in any 60-Hz circuit, including a DC circuit. This AC volt-amp-wattmeter indicates watts of real power; it does not indicate VARS.

In a circuit with voltage and current more or less out of phase, we may state the *apparent power* value, as well as the real power value. Apparent power is measured in volt-ampere units and apparent power is always greater than real power when the voltage and current are out of phase. This topic is considered in detail in the following section because it is very important to the journeyman electrician.

2–6 POWER FACTOR

If the instrument shown in Figure 2–10 is used to measure the voltage, current, and power in a DC circuit, it will be found that the number of watts indicated on the scale is always equal to the product of the voltage and current values. In other words, *all power in a DC circuit is real power, and this real power is equal to the product of volts and amperes.*

Next, if the voltage, current, and power are measured in an AC circuit, it is not necessarily true that the number of watts indicated on the scale is equal to the product of the voltage and current values. In other words, the wattmeter indicates real power, but the real power in an AC circuit may or may not be given by the product of volts and amperes.

The basic power relations in an AC circuit depend upon the load in the circuit, as follows:

1. If the AC circuit has an all-resistive load, such as incandescent lamps, all power in the circuit is real power, and this real power is equal to the product of volts and amperes in the load.

2. If the AC circuit has an all-capacitive load, such as a capacitor, all power in the circuit is reactive power (VARS), and this reactive power is equal to the product of volts and amperes in the load.

3. If the AC circuit has an all-inductive load, such as an ideal inductor, all power in the circuit is reactive power (VARS), and this reactive power is equal to the product of volts and amperes in the load.

4. If the AC circuit has a resistive-capacitive load, or a resistive-inductive load (a frequent practical situation), part of the power in the circuit is real power and

part of the power is reactive power. The proportions of real and reactive power depend on how much the voltage and current are out of phase with each other in the load.

When an AC circuit has a resistive-capacitive (RC load, as in Figure 2–9 (a), an electrician's wattmeter will indicate the amount of real power in the load (resistor R). This value of real power will be less than the product of volts and amperes in the load—*the product of the applied voltage and the current in the circuit is called the number of volt-amperes in the circuit.* Thus, the amount of real power in Figure 2–9 is less than the amount of volt-amperes in the circuit. *A volt-ampere is a power unit; a volt-ampere has some real power and some reactive power. The number of volt-amperes in a circuit is generally called the apparent power in the circuit.*

The relation between real power and apparent power is of basic concern to electricians in many practical situations. This relation is often discussed in terms of the *power factor* of the circuit—particularly by public-utility employees. An electrician can easily determine the power factor of a circuit with a volt-amp-wattmeter, as shown in Figure 2–10. To do so, he makes three measurements and notes the values of the voltage, the current, and the power (real power) in the load. Then, he calculates the power factor from the following fractional expression:

$$\text{Power Factor (PF)} = \frac{\text{Watts of Real Power}}{\text{Volt-Amperes}}$$

As an example, if an electrician finds that a load is drawing 5 amperes at 117 volts, with a real-power reading of 349 watts, the power factor in this circuit is about 0.58, or 58%. We can see that *the real power in a circuit is equal to the number of volt-amperes that are applied, multiplied by the power factor of the load.* The basic power-factor relations in an AC circuit are as follows:

1. If an AC circuit has a purely resistive load, such as an electric heater, all power in that circuit is real power, and the power factor of the circuit is equal to 1 (one).

2. If an AC circuit has a purely capacitive load, such as a capacitor, all power in that circuit is reactive power, and the power factor of the circuit is equal to 0 (zero).

3. If an AC circuit has a purely inductive load, such as an ideal inductor, all power in the circuit is reactive power, and the power factor of the circuit is equal to 0 (zero).

4. If an AC circuit has an RC or an RL load, such as a synchronous motor or an induction motor, part of the power in the circuit is real power, and part of the power in the circuit is reactive power. Thus, the power factor will have a value somewhere in the range from 0 (zero) to 1 (one).

As will be explained in greater detail later, an electrical system can be arranged to operate with a power factor of 1 (100%). A system operates at maximum efficiency when its power factor is 100%.

It is helpful at this point to note and summarize the basic principles of AC circuit action:

1. All AC circuits obey Ohm's law, just as all DC circuits obey Ohm's law.

2. When AC voltage is applied across a resistor, an AC current will flow. The value of this current is given by Ohm's law.

3. AC current flowing through a resistor develops real electrical power. The amount of power in watts is given by the product of the AC voltage and the AC current.

4. When AC voltage is applied across a capacitor, an AC current will flow. A capacitor is somewhat like a resistor in that the capacitor opposes current flow. In other words, the larger the capacitor, the greater is the current flow.

5. AC current flow in a capacitor develops reactive electrical power, and the amount of power in VARs is given by the product of the AC voltage and the AC current.

6. When an AC voltage is applied across a capacitor and a resistor in series, some AC current will flow. The capacitor has *reactance*, and the resistor has *resistance*. The RC load is said to have *impedance*. Impedance opposes current flow, and both real power and reactive power are present in load impedance. The AC source supplies *apparent* power to the load impedance.

2–7 MAGNETISM AND ELECTROMAGNETIC INDUCTION

Almost everyone is familiar with a magnet. However, only an electrician is likely to be familiar with the process of electromagnetic *induction*, shown in Figure 2–11. This is a method of inducing voltage into a coil of wire by moving some magnetic field in and out of the coil. All transformers used by electricians are based on electromagnetic induction. Figure 2–12 also shows an electromagnet being moved in and out of a transformer coil to induce voltage in the coil. The top coil that is connected to the dry cell battery is called the *primary*, and the other coil is called the *secondary*. Note that if the primary coil is placed inside of the secondary coil and the primary circuit to the battery is opened and closed, voltage is induced into the secondary coil. This is the basic transformer induction principle, which is used to step AC voltages up or down.

Figure 2–13 shows the construction of a simple commercial transformer that consists of an iron core with primary and secondary wire windings. If a transformer has more wire turns around its primary than around its secondary winding, it is called a *step-down transformer*. If the transformer has

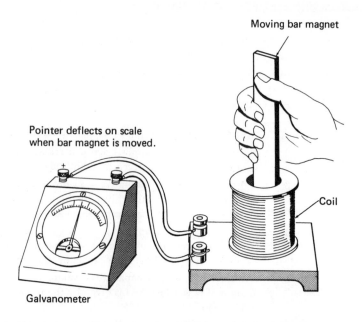

Pointer deflects on scale
when bar magnet is moved.

Moving bar magnet

Coil

Galvanometer

Figure 2-11 Demonstration of electromagnetic induction

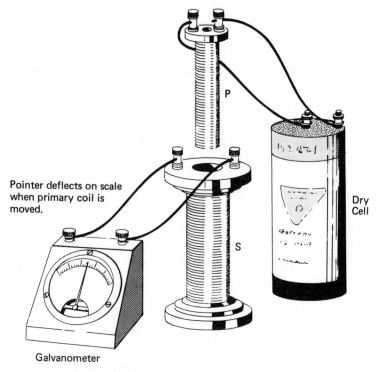

P

Pointer deflects on scale
when primary coil is
moved.

S

Dry
Cell

Galvanometer

Figure 2-12 Another demonstration of electromagnetic induction

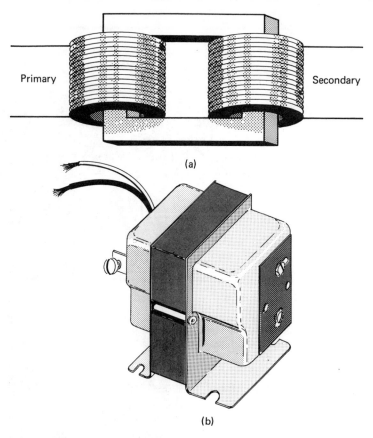

(a)

(b)

Figure 2-13 (a) Construction of a simple transformer (b) A doorbell
ringing transformer

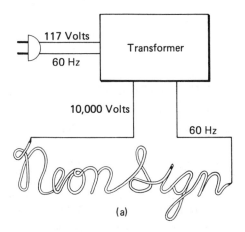

117 Volts

60 Hz

Transformer

10,000 Volts

60 Hz

(a)

Figure 2-14 (a) Step-up transformer used
with a neon sign (Courtesy Westinghouse)

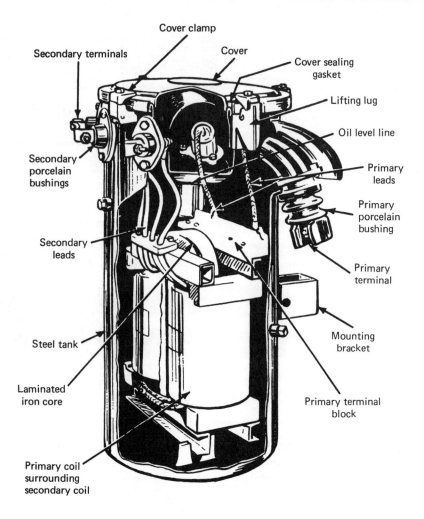

Cover clamp

Secondary terminals

Cover

Cover sealing gasket

Lifting lug

Oil level line

Secondary porcelain bushings

Primary leads

Primary porcelain bushing

Secondary leads

Primary terminal

Steel tank

Mounting bracket

Laminated iron core

Primary terminal block

Primary coil surrounding secondary coil

(b)

Figure 2-14 cont'd. (b) a step-down or distribution transformer used in public utility electrical systems (Courtesy Westinghouse Electric Co.)

more turns on its secondary than on its primary winding, it is called a *step-up transformer*. That is, the difference between the number of primary and secondary wire windings determines the voltage ratio of the primary and secondary.

Figure 2–14 (a) shows a common use for the step-up type of transformer. Here the 117 volts from a 60-Hz line are increased or stepped up to 10,000 volts for the operation of a neon sign. Step-down transformers are widely used to operate doorbells or chimes; they step down 117 volts from a 60-Hz line to approximately 8 volts. They are also used in electric power distribution systems, as depicted in Figure 2–14 (b).

2–8 POWER FACTOR IN AN INDUCTIVE-RESISTIVE CIRCUIT

A coil such as a transformer winding is sometimes called an *inductor*. Electricians are also concerned with coils in motors, and coils in appliances such as dishwashers and clothes dryers. A coil in a motor is called a *motor winding*. A coil used in the automatic operation of appliances is called a *solenoid*. It is evident that a coil in a transformer performs a different task from a coil in a motor. The coil in a dishwasher performs still a different task. It is interesting to note the different tasks performed by various coils, as follows:

1. A solenoid is used in Figure 2–15. It has a sliding iron core or plunger, and it operates basically as a magnet. When alternating current flows through the solenoid, the iron plunger core is pulled into the coil. This motion can be used to open or close valves, to throw switches, and so on.

2. The primary winding of a transformer is used to produce a magnetic field in the iron core of the transformer (Figure 2–13). This magnetic field passes through the secondary winding. In turn, the magnetic field operates to induce an AC voltage in the secondary winding. This induced AC voltage may be higher or lower than the voltage applied to the primary winding.

3. Motor windings can be used to produce magnetic fields or to induce currents in other coils that develop mechanical force. For example, an ordinary induction motor has one fixed winding called a *stator* that produces a magnetic field when alternating current flows through the winding. An induction motor also has a rotating coil arrangement called a *rotor*. The electrical coil arrangement on the iron rotor is called a *squirrel cage*. Because the magnetic field induces alternating current in the squirrel cage, a second magnetic field is produced by the squirrel cage. The first and second magnetic fields develop a force of attraction that keeps the rotor turning around.

In each of the foregoing examples, the coil, inductor, or winding is energized by alternating current. The amount of current that flows in a coil when an AC voltage is applied depends upon the *inductance* of the coil. If a coil has a greater number of turns, it will have a greater value of inductance. Inductance opposes the passage of AC current in somewhat the same way that a capacitor opposes the passage of AC current. An inductor has what is called *inductive reactance*, whereas a capacitor has *capacitive reactance*. We know that reactance is not quite the same as resistance, because AC voltage and current are in phase in a resistive load, whereas AC voltage

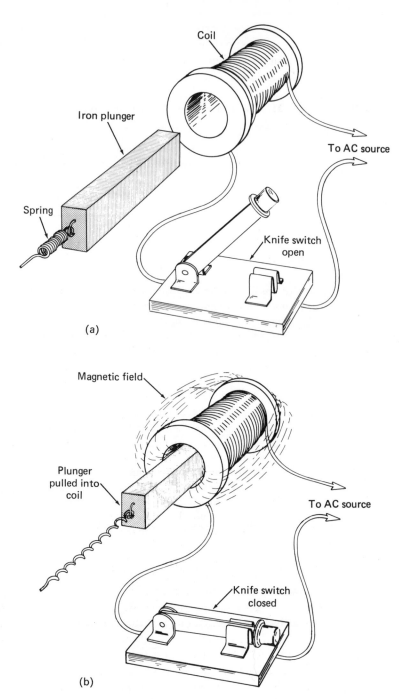

Coil

Iron plunger

Spring

To AC source

Knife switch
open

(a)

Magnetic field

Plunger
pulled into
coil

To AC source

Knife switch
closed

(b)

Figure 2-15 Principle of the solenoid (a) plunger is out of coil with no current flowing (b) plunger is pulled into coil by magnetic action when current is flowing

(a)

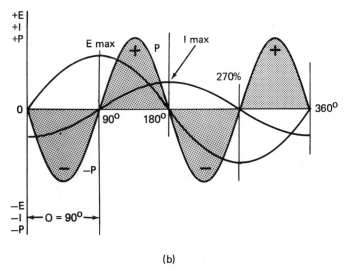

(b)

Figure 2-16 Phase relations in a basic inductive circuit (a) circuit arrangement (b) voltage, current, and power relations

and current are out of phase in a reactive load. An RL load is called an *inductive impedance*, just as an RC load is called a *capacitive impedance*.

From Figure 2–5 we know that the current leads the voltage in a capacitive load impedance. On the other hand, the current lags the voltage in an inductive load impedance, as depicted in Figure 2–16. This is an example of a pure inductive load. The graph shows that the current (I) through an ideal inductor lags the applied voltage by 90°. Recall that the current in an ideal capacitor leads the applied voltage by 90°. As one might expect, the power factor in the circuit of Figure 2–16 is zero.

Note that the unit of inductance is called the henry, as explained in Chapter 1. If a coil has an inductance of 2 henrys, it will pass only half as much AC current as a 1-henry coil with the same value of applied AC voltage. The reactance of a 2-henry coil is twice the reactance of a 1-henry coil. Remember that current will lag voltage by 90° in any size coil, provided that an ideal coil is used. On the other hand, if the coil has resistance, it

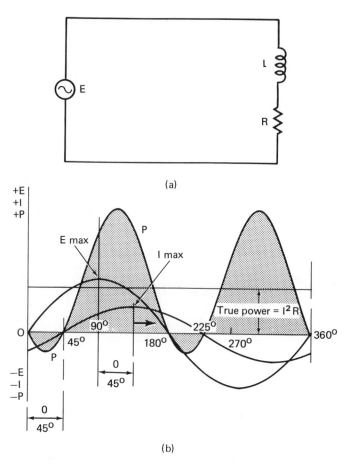

Figure 2-17 Phase relations in an LR circuit (a) circuit arrangement
(b) voltage, current, and power relations

is actually an impedance, and the current-voltage relation is different, as
explained next.

Consider the circuit action of the inductive-resistive (LR) series circuit
shown in Figure 2–17. Here, an AC voltage source E forces AC current to
flow through the load impedance consisting of L and R. We observe from
the graph that the phase relations of voltage, current, and power indicate
more positive power than negative power in the circuit. Observe the difference in the positive and negative shaded areas of the graph. This difference is the result of the fact that the current is lagging the voltage by
less than 90°. In turn, there is both real power and reactive power present.
As in the case of a capacitive-resistive (RC) circuit, the power factor can
be calculated easily from voltage, current, and power measurements with a
volt-amp-wattmeter. The power factor of the inductive-resistive (RL) circuit
is equal to the ratio of watts to volt-amperes.

2–9 CONNECTIONS OF WATTMETER

A wattmeter is used to indicate power values, responding to both voltage and current values in the process. Therefore, the basic wattmeter has four terminals, as shown in Figure 2–18. The voltage terminals of a wattmeter are connected across the load, and the current terminals of the wattmeter are connected in series with the line. Generally, electricians use a wattmeter that has a special adapter cord. If the load is an appliance, such as a blender, toaster, or washer and dryer, the wattmeter can be connected or disconnected quickly. All that is necessary is to unplug the appliance from its outlet (wall socket) and to plug the appliance into the cord adapter of the wattmeter which, in turn, is plugged into the outlet (wall socket). Thereby, the power drawn by the appliance can be measured without disconnecting any wires.

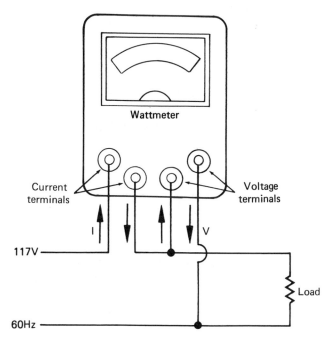

Figure 2-18   Basic wattmeter connections to a load

STUDENT EXERCISE 2-A

The following experiment clearly shows the distinction between volt-amperes, real power, and reactive power.

1.   Obtain a 50-ohm 300-watt resistor, and a 0.15-henry inductor with a current rating of at least 2 amperes. If it is difficult to procure the inductor, a 50-microfarad nonpolarized capacitor may be used instead. A working voltage of 250 volts is ample.

2. Connect the resistor through a volt-amp-wattmeter to a 117-volt, 60-Hz outlet. Note the exact value of the line voltage.

3. Measure the current drawn by the resistor and calculate the number of volt-amperes applied to the resistor.

4. Measure the real power consumed by the resistor and compare this value with the volt-ampere value that you calculated.

5. Disconnect the resistor and substitute the inductor (or capacitor) in its place. Note the exact value of the line voltage.

6. Measure the current drawn by the reactor and calculate the number of volt-amperes applied to the reactor.

7. Measure the real power consumed by the reactor and compare this value with the volt-ampere value that you calculated.

8. Explain why the real power consumed by the resistor showed good agreement with your volt-ampere calculation, whereas the real power consumed by the reactor was much less than the number of volt-amperes that were applied.

QUESTIONS AND EXERCISES

1. What is the difference between AC and DC current?

2. What is the definition of frequency as it applies to alternating current?

3. What is the frequency of most power systems in the United States?

4. What is the meaning of the term "Hertz"?

5. Does Ohm's law apply to AC circuits as well as to DC circuits?

6. Does AC current develop power in a lamp filament?

7. In Figure 2-7, why doesn't the lamp glow when DC voltage is connected to the circuit with the capacitor?

8. What is the phase relationship of current and voltage in a capacitor?

9. What instrument is used to measure power in a circuit in which voltage and current are not in phase?

10. What is meant by apparent power?

11. How is the value of the power factor determined in an electrical circuit?

12. What electrical principle is the basis of transformer operation?

13. What is the relationship between the number of turns in the primary and secondary windings of a step-up transformer?

14. What is the property of an inductor?

15. How is a wattmeter connected to measure power?

16. What is the purpose of a wattmeter?

17. An inductive appliance draws 2 amperes at 120 volts and consumes 200 watts of real power. What is the power factor of the appliance?

3 ELECTRICAL CONDUCTORS

3–1 GENERAL CONSIDERATIONS

Although a conductor is defined as any substance that permits electron flow, the electrician is chiefly concerned with metal wire conductors. Usually electrical wires are made of copper, although aluminum is also used. The diameter of a wire (exclusive of any insulation) is very important. No conductor is perfect and every wire has some resistance, although it may be quite small. Further, every conductor has a certain I^2R power loss that is dissipated and lost in the form of heat. Therefore, to avoid overheating, a wire that has a certain diameter also has a maximum rated current-carrying capacity. This current-carrying capacity rating is called the *ampacity* of the wire. In addition to its basic ampacity value, a wire has various ampacity ratings, depending on the insulation that it may have, as well as its mode of installation. These considerations are explained in detail later.

3–2 WIRE GAGE

Wire diameters, exclusive of insulation, are generally expressed in terms of the American Wire Gage (AWG), also called the Brown & Sharpe Gage (B&S). Actual diameters of copper conductors with gage numbers from 0 to 14 are shown in Figure 3-1. A No. 10 wire has a diameter of 0.102 inch (approximately 0.1 inch). In terms of metric measure, a No. 10 wire has a diameter of 0.259 centimeter (approximately 0.26 centimeter). Relations among centimeters, meters, inches, feet, and yards are as follows:

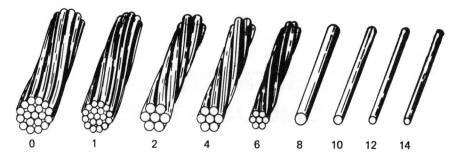

Figure 3-1   Actual diameter sizes of several gage numbers of copper conductors

1 centimeter — 0.3937 inch	1 inch — 2.5400 centimeters (cm.)
1 meter — 3.2808 feet	1 foot — 0.3048 meter
1 meter — 1.0936 yards	1 yard — 0.9144 meter

Electricians usually check the diameter of a wire with a standard wire gage, as shown in Figure 3-2. Note that a wire gage is used on bare wire only. In other words, if the wire has a coating of insulation, this insulation must be removed before the gage of the wire is measured. Note also that the wire is inserted in the slots of the gage—not in the holes. A slot is chosen that provides a fairly snug fit over the wire. Although there are various grades of copper wire on the market, most wiring installations are made with standard annealed copper wire. The resistance of standard annealed copper wire in gage numbers from 0000 to 40 is tabulated in Table 3-1. This table also lists the weights of various sizes of wire, and cross-sectional areas. Diameters are stated in mils (thousandths of an inch).

As would be expected, a small-diameter wire has a greater resistance than a large-diameter wire of the same length. The resistance of a wire is directly related to its ampacity. Electricians, with the exception of motor-repair electricians, seldom measure the resistance of wire directly. Instead, an electrician considers the *voltage drop* that occurs at the end of a line (run of wire).

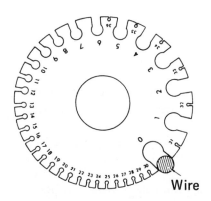

Figure 3-2   A standard wire gage

Table 3-1 Gage numbers, diameters, cross-sectional areas, resistance, and weight of copper wire

(American Wire Gauge — B & S)

Gauge number	Diameter (mils)	Cross section		Ohms per 1000 ft.		Ohms per mile	Pounds per 1000 ft.
		Circular mils	Square inches	25°C. (=77° F.)	65° C. (=149° F.)	25° C. (=77° F.)	
0000	460.0	212,000.0	0.166	0.0500	0.0577	0.264	641.0
000	410.0	168,000.0	.132	.0630	.0727	.333	508.0
00	365.0	133,000.0	.105	.0795	.0917	.420	403.0
0	325.0	106,000.0	.0829	.100	.116	.528	319.0
1	289.0	83,700.0	.0657	.126	.146	.665	253.0
2	258.0	66,400.0	.0521	.159	.184	.839	201.0
3	229.0	52,600.0	.0413	.201	.232	1.061	159.0
4	204.0	41,700.0	.0328	.253	.292	1.335	126.0
5	182.0	33,100.0	.0260	.319	.369	1.685	100.0
6	162.0	26,300.0	.0206	.403	.465	2.13	79.5
7	144.0	20,800.0	.0164	.508	.586	2.68	63.0
8	128.0	16,500.0	.0130	.641	.739	3.38	50.0
9	114.0	13,100.0	.0103	.808	.932	4.27	39.6
10	102.0	10,400.0	.00815	1.02	1.18	5.38	31.4
11	91.0	8,230.0	.00647	1.28	1.48	6.75	24.9
12	81.0	6,530.0	.00513	1.62	1.87	8.55	19.8
13	72.0	5,180.0	.00407	2.04	2.36	10.77	15.7
14	64.0	4,110.0	.00323	2.58	2.97	13.62	12.4
15	57.0	3,260.0	.00256	3.25	3.75	17.16	9.86
16	51.0	2,580.0	.00203	4.09	4.73	21.6	7.82
17	45.0	2,050.0	.00161	5.16	5.96	27.2	6.20
18	40.0	1,620.0	.00128	6.51	7.51	34.4	4.92
19	36.0	1,290.0	.00101	8.21	9.48	43.3	3.90
20	32.0	1.020.0	.000802	10.4	11.9	54.9	3.09
21	28.5	810.0	.000636	13.1	15.1	69.1	2.45
22	25.3	642.0	.000505	16.5	19.0	87.1	1.94
23	22.6	509.0	.000400	20.8	24.0	109.8	1.54
24	20.1	404.0	.000317	26.2	30.2	138.3	1.22
25	17.9	320.0	.000252	33.0	38.1	174.1	0.970
26	15.9	254.0	.000200	41.6	48.0	220.0	0.769
27	14.2	202.0	.000158	52.5	60.6	277.0	0.610
28	12.6	160.0	.000126	66.2	76.4	350.0	0.484
29	11.3	127.0	.0000995	83.4	96.3	440.0	0.384
30	10.0	101.0	.0000789	105.0	121.0	554.0	0.304
31	8.9	79.7	.0000626	133.0	153.0	702.0	0.241
32	8.0	63.2	.0000496	167.0	193.0	882.0	0.191
33	7.1	50.1	.0000394	211.0	243.0	1,114.0	0.152
34	6.3	39.8	.0000312	266.0	307.0	1,404.0	0.120
35	5.6	31.5	.0000248	335.0	387.0	1,769.0	0.0954
36	5.0	25.0	.0000196	423.0	488.0	2,230.0	0.0757
37	4.5	19.8	.0000156	533.0	616.0	2,810.0	0.0600
38	4.0	15.7	.0000123	673.0	776.0	3,550.0	0.0476
39	3.5	12.5	.0000098	848.0	979.0	4,480.0	0.0377
40	3.1	9.9	.0000078	1,070.0	1,230.0	5,650.0	0.299

This voltage drop is directly related to the resistance of the wire. For example, if 117 volts are applied at the beginning of a line, and 97 volts are measured across the terminals of a motor at the end of the line, there is a voltage drop of 20 volts along the line. To avoid this excessive voltage drop, lower-resistance wire must be installed. This topic is explained in greater detail subsequently.

As noted in Table 3-1, the cross-sectional area of a wire is expressed in *circular mils* (CM). Observe that a wire that has a diameter of 1 mil will

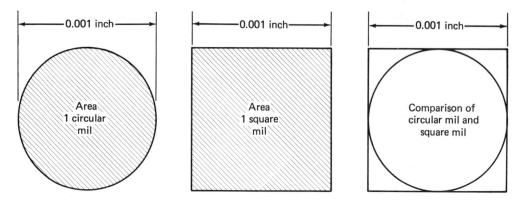

Figure 3-3   Comparison of the circular-mil and square-mil units of area

have a cross-sectional area of 1 circular mil. Or, if a wire has a diameter of 3 mils, its cross-sectional area is 9 CM. This unit of area applies to round wires.

In case the area of a square or rectangular bus is to be stated, the term *square mil* is used. A square wire that is 1 mil on a side has a cross-sectional area of 1 square mil. Or, if a rectangular conductor is 3 mils wide and 2 mils thick, its cross-sectional area is 6 square mils. A comparison of the circular-mil and square-mil units is shown in Figure 3-3.

Note that the area of a *stranded wire* is stated in circular mils. Figure 3-4 shows an example of a 37-strand wire. Each strand has a diameter of 2 mils, or a cross-sectional area of 4 CM. Since there are 37 strands, the total area is equal to 148 CM.

Stranded wire is used in certain applications because it has greater flexibility than a solid wire. An insulated stranded wire is called a *cord.* Cords are used on all portable appliances, lamps, radio and television receivers, and the like. The ampacities, that is the current-carrying capacities, of various wire sizes, are listed in Table 3-2. To estimate the ampacity of stranded wire, see the solid wire gage of the nearest equal cross-sectional area and use

.002 inch
37 strand conductor

Diameter of each strand = .002 inch
Diameter of each strand = 2 mils
Circular mil area of each strand = D^2 = 4 CM
Total CM area of conductor = 4 x 37 = 148 CM

Figure 3-4   Cross-sectional area of a stranded wire

Table 3-2 Ampacities of various sizes of copper wire

Wire size	In conduit or cable		In free air		Weather-proof wire
	Type RHW* THW*	Type TW, R*	Type RHW* THW*	Type TW, R*	
14	15	15	20	20	30
12	20	20	25	25	40
10	30	30	40	40	55
8	45	40	65	55	70
6	65	55	95	80	100
4	85	70	125	105	130
3	100	80	145	120	150
2	115	95	170	140	175
1	130	110	195	165	205
0	150	125	230	195	235
00	175	145	265	225	275
000	200	165	310	260	320

*Types "RHW," "THW," "TW," or "R" are identified by markings on outer cover

that ampacity value. Gage numbers are assigned to stranded wire on this basis. Consequently, a stranded wire that has the same gage number as a solid wire is somewhat larger in diameter.

To an increasing extent aluminum wire is being used in household wiring systems. Aluminum wire is easier to work with than copper. Although it does not conduct as well as copper, it weighs less and costs considerably less. Because of its higher resistance, aluminum wire must be one size larger than copper in a given circuit. For example, if No. 14 copper wire has correct ampacity for a certain circuit, the electrician should use No. 13 aluminum wire in the same circuit. Electricians generally use even sizes of wire. Therefore, a No. 12 aluminum wire is used instead of No. 14 copper wire, in practice.

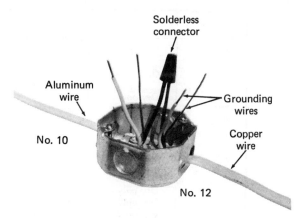

Figure 3-5 Splicing aluminum to copper wire with solderless connectors

Aluminum wire can be spliced to copper wire with conventional solderless connectors, as shown in Figure 3-5. Some fuse panels and circuit-breaker panels are not approved for aluminum wire. In such a case, the electrician splices onto the aluminum wire a short length of copper wire of the same ampacity before making the terminal connection. If "Al-Cu" is printed on a wiring device, the electrician knows that it is approved for both aluminum and copper wires. However, several approved devices do not have this marking. In case of doubt, use a short copper wire linkage to ensure that the finished job will pass inspection.

3—3 INSULATED WIRES

House wiring systems generally use plastic-insulated wire. Thermoplastic-insulated wire is called *Type T*. It is approved for use only in dry locations such as living rooms, bedrooms, and so on. *Type TW* wire also has thermoplastic insulation but is approved for installation in either wet or dry locations. For example, Type TW wire may be used in washrooms and garages, embedded in concrete, or buried underground. A third kind of wire, *Type THW* is also approved for installation in wet locations but has a higher temperature rating, as seen in Table 3-3. Therefore, Type THW wire has a higher ampacity rating than TW.

Residential and similar wiring systems may also employ *Type THHN* or Type THWN wire. Less thermoplastic insulation is used on these, but an outer covering of nylon is provided. Nylon is a very good insulator and has a tougher texture than thermoplastic. Therefore, THHN or THWN conductors have a smaller outside diameter than T, TW, or THW conductors.

Table 3-3   Rated temperature limits for various conductors
(National Electrical Code)

Type	Locations	Sizes	Temperature limits
RHH	Dry only	Nos. 14, 12, 10 No. 8 and larger	75%C or 167%F 90%C or 194%F
RHW	Dry or wet	All sizes	75%C or 167%F
T	Dry only	All sizes	60%C or 140%F
TW	Dry or wet	All sizes	60%C or 140%F
THW	Dry or wet	All sizes	75%C or 167%F
THWN	Dry or wet	All sizes	75%C or 167%F
THHN	Dry only	Nos. 14, 12, 10 No. 8 and larger	75%C or 167%F 90%C or 194%F
XHHW	Dry only Dry only Wet only	Nos. 14, 12, 10 No. 8 and larger All sizes	75%C or 167%F 90%C or 194%F 75%C or 167%F

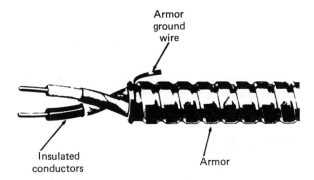

Figure 3-6 Appearance of armored cable (Courtesy General Electric Co.)

This smaller-size advantage can be a deciding factor when certain types of wiring systems are being planned.

A somewhat similar wire is called Type XHHW. This employs a special polyethylene insulation, instead of thermoplastic insulation, and has superior qualities. Type XHHW wire is approved for use in either wet or dry locations. Figure 3-9 illustrates these varieties.

Although less widely used than the types described above, rubber-insulated wire has not become obsolete. To avoid a chemical reaction with the rubber, copper wire is always tinned when rubber insulation is used. Tinning the copper also makes it easier to solder joints in some installations that prohibit solderless connectors. The rubber insulation is covered by a fabric braid that is impregnated with fire-resistant and moisture-repellent substances. *Type R* wire has been superseded by *Type RHW* which is approved for installation in either wet or dry locations. *Type RHH* has superior heat-resistant insulation and has a higher ampacity rating. However, it is approved for use *only* in dry locations.

A cable is a stranded wire that is heavier than No. 4 gage. Electricians also refer to grouped wires as *cable*. [Cables have a practical advantage in that they speed up a wiring job.] Cable is widely used in residential and commercial wiring systems, as explained in detail in later chapters. A 12-2 cable contains two No. 12-gage wires. A 14-3 cable contains three No. 14-gage wires. As seen in the photos of aluminum wire, one bare (uninsulated) wire is usually included. This bare wire is used to ground appliance or device housings, as shown in Figure 3-5. Appliance housings are grounded, via the bare wire, to minimize the possibility of shock if the appliance becomes defective.

Nonmetallic sheathed cable is manufactured in *NM* and *NMC* types. The NM cable is approved only for dry locations, but Type NMC may be used in either wet or dry locations. Type NM cable may consist of rubber-insulated wires that are covered with an additional spiral layer of tough paper. The space between the paper coverings is filled with impregnated jute. Finally, the cable is jacketed with an impregnated fibrous substance. Type NM cable may also use thermoplastic insulation instead of rubber. This type may or may not have spiral paper layers. For wet locations, Type NMC cable is required. This design features a solid plastic jacket that provides immunity to moisture. Note that *NM-G* cable is provided with a grounding wire.

3–4 ARMORED CABLE

Armored cable (Figure 3-6) is quite rugged. The flexible spiral armor is made from galvanized steel. An *armored cable* usually has two wires, distinguished by black insulation and white insulation in American practice. However, three-wire armored cable is also available; the third wire has red insulation. This is the same color-coding convention used with other electrical wires. Typically, the grounded neutral wire is always white; the circuit wires are black, red, blue, or yellow, depending on the number of circuits that are used in the system. Note that the armor is always grounded, and that inside the armor there is a bare copper wire or strip to ensure effective grounding. A grounding wire is required.

The two or more wires in an armored cable may be rubber-insulated or thermoplastic-insulated. *Type AC* cable has rubber insulation and *Type ACT* cable has thermoplastic insulation. Electricians generally refer to both types as *BX cable*. Note that the cable wires shown in the figure are spiral-wrapped with a layer of impregnated tough paper. Armored cable is approved for installation only in *dry* locations. Note that flexible conduit (also called Greenfield) is essentially the outer sheath of an armored cable. In other words, the wires are pulled into the flexible conduit after it has been installed. A grounding wire is required. It may be bare or have green-colored insulation, or a wrapping of green with a yellow stripe. Flexible conduit and armored cable have an outer diameter slightly greater than 3/4 inch. Greenfield is used chiefly in modernizing old wiring systems, as explained in detail in a following chapter.

Armored cable is available in 50- to 200-foot rolls in a wide range of wire sizes. In wet locations, lead-covered armored cable is required. This is designed with a continuous lead sheath over the wires, so moisture that penetrates the armor cannot reach the wires. This type of armored cable is called *ACL*. Armored service-entrance* cable is called *ASE*. It is made to withstand moisture and to conduct the heavy current values of service installations. Armored cable is more difficult to install than is open wiring and is more costly. With the development of new types of insulation for wiring, armored cable is being installed in fewer electrical systems than in

*A service entrance consists of the conductors and equipment used to connect a main line to the area to be served, such as a residence, office, or factory.

the past. Open wiring consists of cable which is secured to wood structures by staples, as detailed in following chapters. Armored cable has been installed extensively in older buildings.

3–5 OUT-OF-DOORS WEATHERPROOF WIRE

Weatherproof wire is specially designed for outdoor applications and is not approved for indoor use. However, various types of approved indoor wire, such as thermoplastic or rubber-coated types, may be installed outdoors. Weatherproof wire consists of a copper or aluminum conductor with a neoprene plastic coating or jacket. Coated wire is in the same basic category as a bare wire. When a given gage of wire is used in an outdoor electrical circuit or "run," it is accorded a higher ampacity rating than when it is used in an indoor run.

3–6 TYPES OF FLEXIBLE CORDS

As noted before, the line connections to lamps or portable appliances (appliance cords) must be made with flexible cords. The *SP* cord type consists of 40 strands of copper wire with rubber insulation. *Type SPT cord* has thermoplastic insulation with 40 strands. Cord gage numbers range from No. 18 to No. 12. Moderately heavy (high-current) appliances, such as washing machines and vacuum cleaners, utilize *Type S cord* that is more extensively insulated. A typical washer-dryer will draw 20 or 25 amperes. Type S cord is manufactured in sizes as large as No. 2. It has the sturdiest construction and is used in both industrial and residential applications. *Type SJ cord* has a lighter construction and is used to a limited extent in industrial applications. *Type SV* is approved for use with vacuum cleaners in residential use only.

To minimize the hazard of shock from defects that may occur in some electric typewriters, appliances, laundry dryers, and the like, a three-conductor cord is used. This is shown in Figure 3-7(a). Here a green grounding wire is included with the two circuit wires, and the plug cap has three prongs instead of two. (In the case of three-conductor cords, green is the standard color code for the grounding wire). A plug cap terminates a cord, and has metal prongs for insertion into an outlet. The grounding wire is connected to the metal housing of the appliance and to the grounding wire in the outlet box. This ensures that the appliance will remain "cold," even if a short circuit develops between the electrical parts in the appliance and its metal housing. An appliance is said to be "cold" if its housing is at ground potential. Of course, a fuse will blow or a circuit breaker will trip in this situation. Note that a three-prong plug cap will fit into a special mating receptacle. This is called a grounding receptacle. Figure 3-7(b) shows the arrangement of a grounding receptacle.

Heater cords used on portable heaters, electric irons, hot plates, toasters,

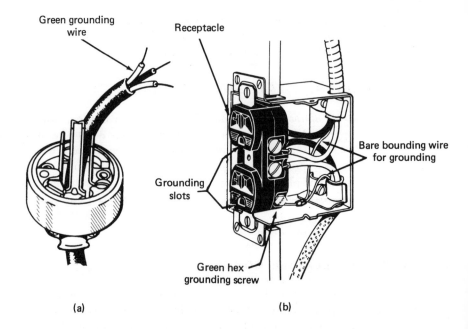

Figure 3-7 (a) Three-conductor cord with three-prong plug cap (b) arrangement of a grounding receptacle

and similar appliances may have asbestos insulation plus an outer braid of cotton or rayon. This is called *HPD cord*. Some heater cords have an outer jacket of rubber and are termed *Type HSJ*. *Type HPN cord* is also in wide use and employs neoprene insulation. So-called *fixture wire* is used only inside of lighting fixtures. This is a heat-resistant type of cord. *Type AF* fixture wire has asbestos insulation; *Type CF* has cotton insulation; *Type SF* has silicone insulation. Journeyman electricians soon become familiar with other types of flexible cords that are used in special applications.

STUDENT EXERCISE 3-A

This exercise provides familiarity with typical electrical cable and wires.

1.   Obtain samples of Type NM 14-2 nonmetallic sheathed cable, 14-2 G nonmetallic sheathed cable, 12 Type TW wire, and 2/7 Type TW Oil and Moisture Resistant stranded cable. Note that G denotes provision of a grounding wire.

2.   Remove the insulation from the end of each wire and cable with a pocket knife, being careful not to nick the wire conductor.

3.   Observe that the Type NM 14-2 cable contains two No. 14 insulated wires, whereas the Type 14-2 G cable contains a grounding wire in addition to the pair of No. 14 insulated wires.

4. Note that the 2/7 TW stranded cable is equivalent to a solid No. 2 conductor, but that it is made up from 7 smaller conductors.

5. Using a standard wire gage, measure the size of each conductor in the sample cables and wires. Check to determine whether the grounding wire in the 14-2 G cable is the same size as the insulated conductors.

6. Observe the color coding of the insulated wires in the cables.

7. On the basis of your measurements, calculate the number of circular mils in the sample of 2/7 TW stranded cable. Compare the number of circular mils specified in a wire table for No. 2 wire with the number of circular mils that you calculated.

STUDENT EXERCISE 3-B

This exercise provides a practice in calculating cable requirements. See the wiring layouts in Figures 3-8 and estimate approximately:

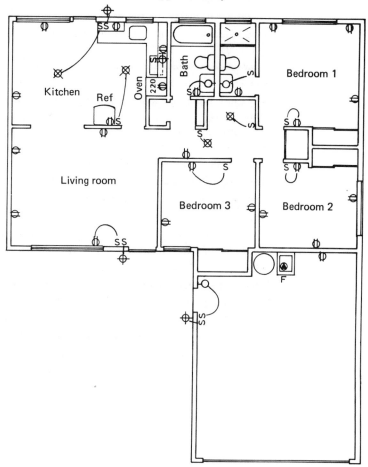

Figure 3-8 Electrical wiring layout

1.   The number of feet of cable required for each circuit
2.   The total number of feet of cable required for the 120-volt circuits
3.   The total number of feet of cable required for the 240-volt circuits
4.   The total number of feet of cable required for the complete wiring system.

Also determine or estimate the following requirements:

5.   The number of 120-volt duplex outlets required
6.   The number of toggle switches that will be necessary
7.   The number of 240-volt outlets required
8.   The total number of outlet and switch boxes required for the complete wiring system

QUESTIONS AND EXERCISES

1.   What is ampacity?
2.   How can you determine the gage of a certain piece of copper wire?
3.   In what unit of measurement is the cross-sectional area of a conductor usually expressed?
4.   What name or names are given to stranded wire?
5.   How can you estimate the ampacity of a stranded wire?
6.   Why must aluminum wire be one gage larger than copper wire for a similar application?
7.   How does an electrician know that an electrical device is approved for both aluminum and copper wire?
8.   What use limits are placed on Type T wire?
9.   What are the applications of type THW wire?
10.   What is the practical advantage of using electric cables?
11.   What is the advantage of armored cable over conventional cable?
12.   What two letters do electricians generally use to denote armored cable?
13.   What type of armored cable is required wherever there is moisture?
14.   What is meant by the term Greenfield?
15.   What is the advantage of type S cord over SV cord?
16.   Explain the purpose of a grounding wire.
17.   What color is standard for a grounding wire in a three-conductor cord?
18.   How much current does a typical washer-dryer draw?
19.   What does the letter G denote in cable identification?
20.   What is the purpose of three-conductor cord?

4 ELECTRICAL DEVICES

4–1 OUTLETS

An *outlet* is a device in a wiring system from which current may be taken to supply lamps, heaters, fixtures, and so on. In some cases, an outlet is simply a connection point. It is one of the devices in a line that does not consume power. Strictly speaking, a *device* carries current, but does not consume power. Thus, a switch is a true device. However, a receptacle for an outlet is mounted in the same kind of metal outlet box as a lighting switch. The technical distinction between outlets and devices tends to be disregarded in practice. Until the 1970's the basic wall outlet was provided with two terminals connected to a line. However, due to shock hazards from defective appliances and lamps, later electrical standards required that the basic wall outlet provide three terminals. This third terminal is connected to a ground in the wiring system, as explained later in detail.

Figure 4–1(a) shows a modern *single-receptacle outlet*. This type of outlet serves one fixture only. *Duplex receptacle outlets*, such as depicted in Fig. 4–1(b), are more widely used because they accommodate the connecting plugs for two appliances at a single outlet box. However, a single receptacle outlet may be combined with a toggle switch, as in Figure 4–1(c). A *clock-hanger outlet*, is set into the wall and recessed so that an electric clock can be hung flush against the wall. *Electric range receptacle outlets*, illustrated in Figure 4–1(d), operate at 240 volts and have a different plug arrangement, so an appliance that takes only 120 volts cannot be accidentally connected to the 240-volt line. Combination outlets that supply power as well as providing an antenna receptacle for TV are in fairly wide use. *Weatherproof receptacle outlets*, which are enclosed, are used in outdoor wiring systems.

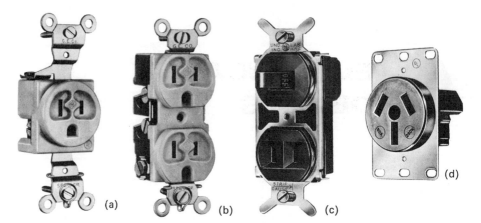

Figure 4-1   (a) Single receptacle outlet (Courtesy General Electric Co.) (b) duplex receptacle outlet (Courtesy General Electric Co.) (c) single receptacle outlet with toggle switch (Courtesy Underwriters' Laboratories Inc.) (d) electric range outlet (Courtesy Underwriters' Laboratories Inc.)

4–2 SWITCHES AND SWITCHING METHODS

Various types of switches are used in wiring systems. One of the most widely used types is the *toggle switch* shown in Figure 4–2(a). The spring-loaded design makes a thud when the switch is thrown. One the other hand, a mercury contact switch design is noiseless. A delay type of toggle switch that turns on instantly is also available. However, when thrown to its off position, the *delay switch* does not open for several seconds, say 15 seconds. This enables a person to leave a room before the lights go off. Some designs have an adjustable time delay.

Two toggle switches may be mounted in the same switch box. This is called a *two-gang switch*; it controls two separate lights, for example. *Push-button switches*, as shown in Figure 4–2(b), are noiseless, but do not employ mercury contacts. A special mechanical design is employed which is less costly than the mercury-contact arrangement.

Several types of switches have pilot lights that glow whenever the switch is turned on. One design has transparent handles that contain neon lights. The neon pilot light has a comparatively long life. Another type of switch contains a dimming device. It is called a *three-position light-dimming switch*, has high, low, and *off* positions. A key-operated switch, shown in Figure 4–2(c), can be turned *on* or *off* only with a proper key. This type of switch is often installed where children are likely to play with the switches.

Figure 4–3(a) shows a *weatherproof switch*; the plate (front cover) is mounted on a weatherproof switch box with a gasket. A *photoelectric control switch* may

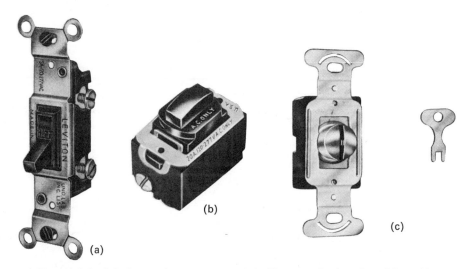

Figure 4-2   (a) Conventional toggle switch (Courtesy Leviton Inc. (b) pushbutton switch (c) a key-operated switch (Courtesy General Electric Company)

also be installed outdoors in a weatherproof switch box with a gasket. A photoelectric-type switch automatically turns on lights at sunset and turns them off at sunrise. A *rotary dimmer switch* is shown in Figure 4–3(b). This switch is used with tungsten-filament lamps and provides a continuous range of brightness control. A *door switch*, turns lights on and off automatically

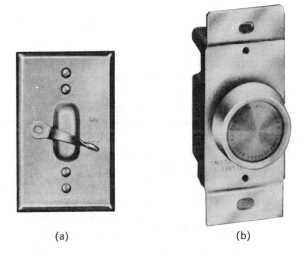

(a)   (b)

Figure 4-3   (a) Weatherproof switch   (b) rotary dimmer switch   (Courtesy General Electric Co.)

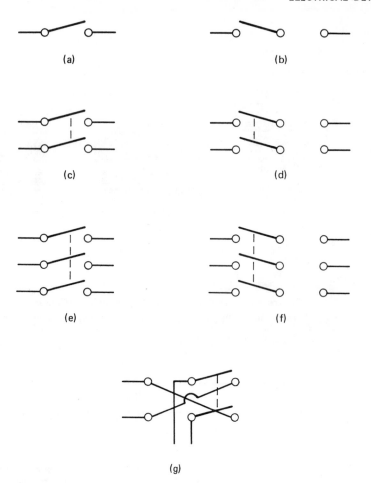

Figure 4-4 Basic switch arrangements (a) single-pole single-throw (SPST) (b) single-pole double-throw (SPDT) (c) double-pole single-throw (DPST) (d) double-pole double-throw (DPDT) (e) triple-pole single-throw (f) triple-pole double-throw (g) polarity reversing

whenever a door is opened or closed. A small door switch is used to turn on the light in electric refrigerators. A *safety switch* is used in combination with a service panel.* Note that a safety switch differs from the older knife switch in that the electrically "hot" portion of the arrangement cannot be accidentally touched by a person while operating the switch.

A *service* is defined as the conductors and equipment for supplying electrical energy from the mains, feeder, or public-utility transformer to the area

---

*A service panel is part of the equipment included in a service entrance, as discussed later.

that is served with electric power. An important unit in a service is the *service switch*. Service switches are designed as safety switches.

Basic types of switches used in wiring systems are depicted in Figure 4–4. A *single-pole single-throw* (SPST) switch is an on-off device that operates in one side of a line. Next, a *single-pole double-throw* (SPDT) switch permits a conductor to be connected to either of two other conductors. A *double-pole single-throw* (DPST) switch is an on-off device that operates in both sides of a line. Again, a *double-pole double-throw* (DPDT) switch permits a pair of conductors to be connected to either of two other pairs of conductors. A *triple-pole single-throw* switch is an on-off device that operates in each lead of a three-wire circuit. This topic is explained in greater detail under three-phase circuits in a following chapter. Again, a *triple-pole double-throw* switch permits a group of three conductors to be connected to either of two other groups of three conductors. Note that a polarity-reversing switch is basically a DPDT switch connected to a pair of conductors in such manner that the outgoing conductors can be electrically transposed (reversed) with respect to the incoming conductors, by throwing the switch.

A basic switching circuit for a lighting fixture is shown in Figure 4–5. It makes use of a toggle switch with SPST construction. This switch provides ON-OFF operation in the "hot" side of the line. Leaving the fixture electrically "dead" when the switch is in OFF position.

Now consider the arrangement for a *two-gang switch*—that is, two switches installed in one box under one faceplate, with each switch controlling a separate outlet, as pictured in Figure 4–6. Note that the switch loop is taken from outlet box No. 2. Four wire splices are required in box No. 2, and two wire splices in box No. 1. As in the previous example, all switching is done in the "hot" wire. Observe that switch No. 1 controls outlet No. 1, and switch No. 2 controls outlet No. 2.

It is often desirable or necessary to control one or more lamps from two different locations. A long corridor, a stairway, or a room with two separate entrances will require switch facilities at two locations for a lighting fixture. This type of switching action is provided by a pair of *3-way switches*. Refer to Figure 4-7 and observe that a 3-way switch is basically an SPDT switch. In turn, there are three wires connected to a 3-way switch. Note that the single terminal, shown at the bottom of the base in the diagram, is called a *hinge point*. A black oxide screw may be provided to identify the hinge-point terminal. Observe that the "hot" wire is connected to the hinge point in the switch-leg portion of the circuit. The hinge point in the return-leg portion of the circuit is connected to the "hot" terminal of the lamp fixture. The other two wires that run between the 3-way switches are called *travelers*.

Note that the path for current flow can be traced from the "hot" supply wire in the outlet box to the hinge point of switch No. 1, then through the right-hand traveler terminal, and on through the upper traveler to the left-hand traveler terminal of switch No. 2, where the circuit is "open". Since

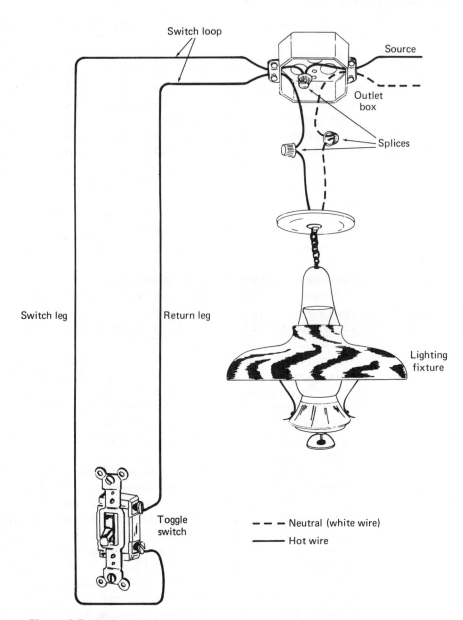

Figure 4-5 Basic switching circuit

both switches are in the same position and there is no connection from the left-hand traveler terminal to the hinge post, the lamp is OFF. However, if the lever of switch No. 2 is thrown to its other position, current can flow from the left-hand traveler to the hinge point, through the return leg to the lamp socket, on through the lamp filament to the neutral wire, and the lamp is

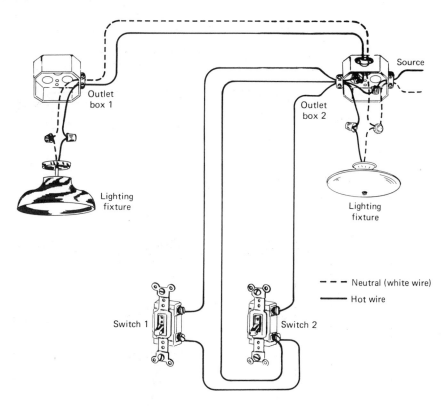

Figure 4-6   A two-gang switch wired to turn two fixtures on or off independently.

ON. Next, suppose that the switch levers are returned to their original positions, and the lamp is OFF. If the lever of switch No. 1 is thrown to its other position, current flow from the hinge point is to the left-hand terminal, through the bottom traveler to the right-hand terminal of switch No. 2 and to the hinge point, through the return leg to the lamp socket, and through the lamp filament to the neutral wire, causing the lamp to go ON. In this way either of the 3-way switches can be thrown to its other position to change the lamp from on to off, or from off to on, as needed.

When a fixture is to be turned on or off from three separate locations, a 4-way switch must be used as well as two 3-way switches. Figure 4–8 shows a wiring diagram for this arrangement. Note that a 4-way switch is basically a transposition or polarity-reversing switch, a form of double-pole double-throw switch. If more than three locations are required, the electrician installs an additional 4-way switch at each additional location. Each additional 4-way switch is connected into the circuit in the same way as shown for the 4-way switch in the center of Figure 4–8. Observe that corresponding wires are denoted by corresponding letters in the diagram.

It is essential to follow the details of the wiring diagram in Figure 4–8.

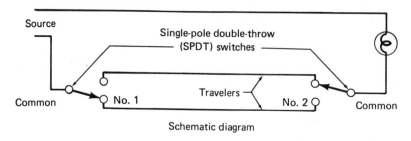

Schematic diagram

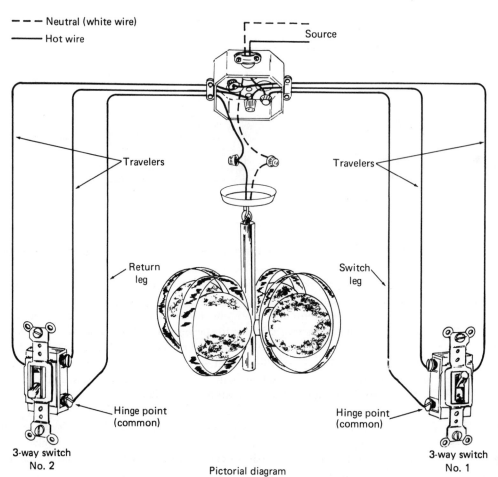

Pictorial diagram

Figure 4-7 A three-way switching circuit

In other words, the electrician must be careful to connect the black, white, and red wires as shown. Note that the switches have light- and dark-colored terminals; the terminals marked A and B in the diagram are the light-colored terminals to which red and white wires must be connected. Terminal

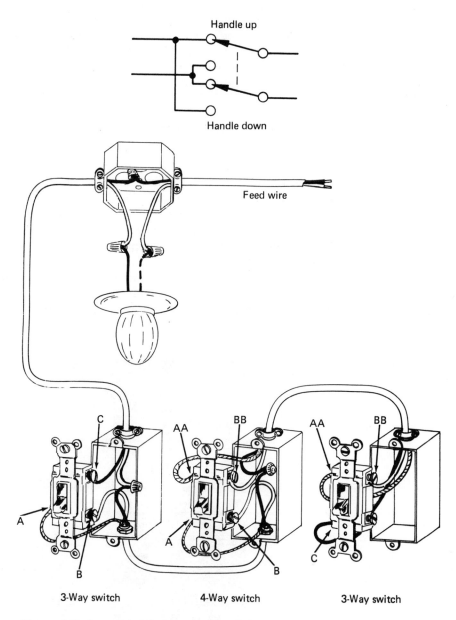

Handle up

Handle down

Feed wire

C

AA

BB

AA

BB

A

A

B

B

C

3-Way switch   4-Way switch   3-Way switch

Figure 4-8 Four-way switch used with two three-way switches

C is the dark-colored terminal to which the black wire is connected. Some switches may have the terminals located somewhat differently than indicated in the diagram. Neverthless, the light and dark colors of the terminals provide a reliable guide for the electrician. In the circuit of Figure 4–7, note also that, to keep the color coding consistent, the white wire from the switches must be painted black, both at the switches and at the fixture. Consistent color coding is not required by all codes and regulations. However, in those localities that require it, the electrician has the responsibility of maintaining consistent color coding, in order that the completed wiring installation will pass inspection.

4–3 INSTALLATION OF A TYPICAL SWITCH BOX

Now let us follow the steps that are taken in the installation of a switch box, as shown in Figure 4–9. In this example, a metal box is mounted with two nails onto an upright wall stud. The box is positioned on the stud so that its edges will be flush with the surface of the panelboard that is nailed to the studding when the wall is finished. Next, the cable is inserted into the box and clamped in place. The clamping procedure is pictured in Figure 4–10. Connections are made to the switch, and the switch is secured in place by two machine screws. After the panelboard is nailed to the studding, a switch plate is secured to the box with two machine screws. Installation of various receptacles is explained in greater detail later. Typical machine screws are illustrated in Figure 4–11.

Figure 4-9 How a junction box is nailed to a stud

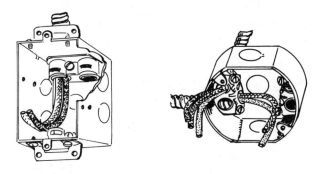

Figure 4-10 Conductors are often anchored in boxes by means of built-in cable clamps

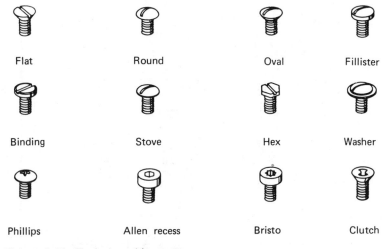

Flat	Round	Oval	Fillister
Binding	Stove	Hex	Washer
Phillips	Allen recess	Bristo	Clutch

Figure 4-11 Typical machine screws

4–4 OVERLOAD PROTECTIVE DEVICES

Overload protective devices must be provided for each part of a wiring system. In addition, protection must be provided for certain kinds of load equipment. These overload devices are built into residential service panels, such as shown in Figure 4–12. A service panel contains several *circuit breakers* or several fuses. These are also called *overcurrent devices*. When an excessive current demand is suddenly made on the electrical system, one or more of the overcurrent devices will act to open the circuit(s). If a fuse has blown because of too great a power demand, it must be replaced. Of course, one must first clear the fault that caused the circuit(s) to open automatically.

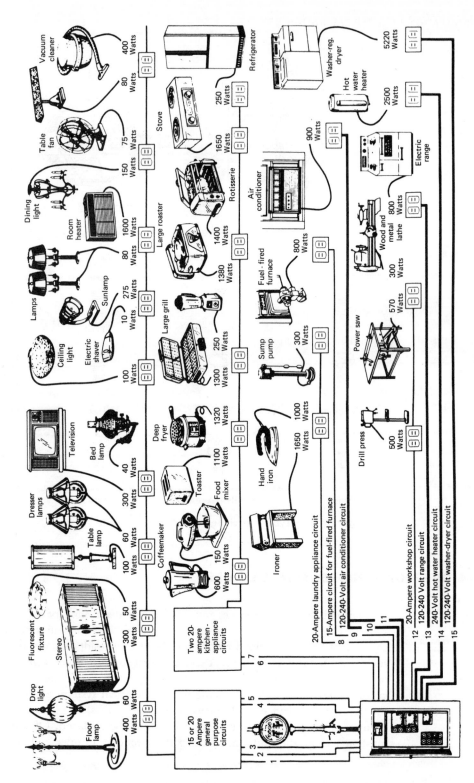

Figure 4-12 Examples of circuits fused for various maximum current demands

Otherwise it will be impossible to keep the breaker closed or to keep a fuse intact.

Figure 4–12 shows some examples of circuits that have been fused for their maximum current demands. No branch circuit in this system has a current demand of more than 20 amperes. Each branch circuit is protected by an overcurrent device in the service panel. Note that appliances like the range, the washer-dryer, and the water heater that operate at 240 volts draw half as much current as an equivalent appliance operated at 120 volts. This fact is shown in Figure 4–13. Here, two ideal motors, each rated at 1 horse-power (746 watts), are operated on different line voltages. Note that the 120-volt motor draws 6.21 amps to develop 1 HP, whereas the 240-volt motor draws 3.11 amps to develop 1 HP. Overcurrent devices are connected into the "hot" side of a circuit only. In other words, a 120-volt circuit has one fuse or circuit breaker; a 240-volt circuit has two fuses or circuit breakers. A separate three-wire 240-volt branch circuit is used for an electric range, water heater, dryer, and the larger types of air conditioners. A 100-ampere service can supply power for a range, heater, and ordinary dryer.

A range is operated at 240 volts on high heat and at 120 volts on low heat. It requires a three-wire 50-ampere branch circuit of its own from the main service panel. The heater element for a dryer is operated at 240 volts. However, a dryer motor and light operate at 120 volts. An individual 30-ampere circuit is required to operate the average dryer unit. A high-speed dryer requires an individual 50-ampere circuit. Note that a temporary overload occurs when a motor starts up, because its starting current is about three times greater than its running current. Special time-delay fuses are utilized in a motor circuit. Circuit breakers are designed to withstand temporary overloads without tripping.

Any motor draws a comparatively large starting current, because the first surge of starting current is limited only by the winding resistance of the coils in the motor. However, as soon as the motor armature begins to rotate, the motor develops generator action. This generator action produces a counter emf (CEMF) that opposes the voltage applied to the motor. In effect, the value of the applied voltage has been reduced or, from another

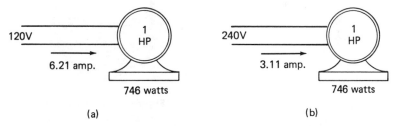

(a)   (b)

Figure 4-13 Examples of two motors that are operated at different line voltages with the same power. (a) 120-volt motor draws 6.21 ampere (b) 240-volt motor draws 3.11 ampere

viewpoint, the internal resistance of the motor has been increased. The result is to reduce the current demand of the motor. As the motor armature comes up to full speed, the value of the CEMF increases to its maximum value. Now, the current demand of the motor is at its minimum value. This is the rated current demand of the motor, provided that it is normally loaded. Typically, the rated current demand is about one-third of its starting current demand.

4–5 TYPES OF FUSES

A 15-ampere plug-type fuse is shown in Figure 4–14. It has a transparent top so you can see if the fuse element is blown. This type of fuse is available in 10-, 15-, 20-, and 25-ampere ratings. They are used only in 120-volt circuits, or in 240-volt circuits where there are 120 volts between each hot wire and ground. An example of a four-wire system in which there is 120 volts between each hot wire (A, B, or C) and ground is shown in Figure 4–15. Note that the neutral wire is at ground potential. This is called an arrangement of three separate 120-volt single-phase feeders for lighting; it is derived from a 3-phase 120/208-volt 3-phase service. The distinction between single-phase and three-phase lines is explained in a later chapter.

When a fuse blows, it is advisable to unplug all of the lights and appliances in that circuit. In other words, all lights and appliances that will not work should be unplugged and switched out of the circuit. Then when a replacement fuse is inserted, it will probably not blow. Then, to find the overload, turn on each lamp and appliance and leave them on until the new replacement fuse blows. This procedure quickly shows the source of the overload. Sometimes a replacement fuse will blow as soon as it is inserted, although all of the circuit loads have been unplugged or switched off. When this happens,

Figure 4-14 A 15-ampere plug-type fuse with Edison base and a window top

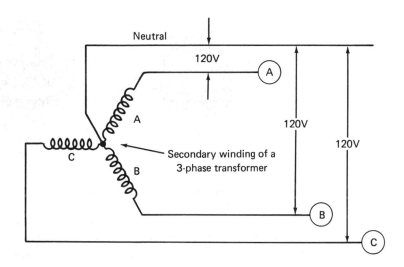

Figure 4-15 Example of a four-wire system with 120 volts between each hot wire and ground

there is a defect in the wiring system that must be corrected. The correct way to troubleshoot wiring systems is explained in detail later.

As noted before, an electric motor draws starting current that is greater than the operating current. This starting load surge does not endanger the wiring system, but a continuous load of this value would be dangerous. A standard fuse would blow under a starting load in a motor circuit. So for the starting load, the electrician installs another kind of fuse that permits a starting load to be drawn for a short time without blowing the fuse. This fuse is called a *time-lag,* time-delay, or *slow-blow fuse.*

Figure 4–16 shows a commercial version of the time-lag fuse called the Fusetron. This has a two-dual element that consists of a fusible wire element and a small solder cup with a spring attached to the fusible element.

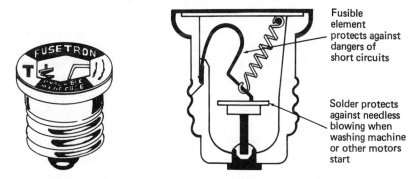

Fusible element protects against dangers of short circuits

Solder protects against needless blowing when washing machine or other motors start

Figure 4-16 A typical time-lag fuse

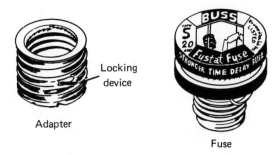

Figure 4-17 A Type S, or Fustat ® type of fuse

Any fuse tends to run hot. Under continuing overload, the fuse element heats so much that the solder melts, and the spring action opens the circuit. Under a temporary overload, however, the solder does not melt. If a short-circuit fault occurs, the fusible element blows.

Fuses like those shown in Figures 4–14 and 4–16 have Edison bases. To prevent accidental (or deliberate) replacement of low-current fuses by high-current fuses, *Type S* fuses or Fustats ® have been developed. This type of fuse, shown in Figure 4–18, has special thread sizes and cannot be used in an Edison fuse socket unless a mating adapter is utilized. Note that this adapter has a locking arrangement, so that once the adapter has been inserted in the Edison socket it cannot be readily removed. In turn, the Type S fuse can be inserted into the adapter and can be easily removed in case replacement is required. The inside diameter of the adapter is such that the fuse cannot be defeated (short-circuited by inserting an object such as a penny). Note that a Fustat does not defeat the purpose of a conventional fuse, because a Fustat permits an over current demand only for a short time.

Circuits that draw more than 30 amperes use cartridge-type fuses. Even if less than 30 amperes is needed, both the 250-volt and 600-volt circuits often employ cartridge fuses. Figure 4–18 shows two ferrule-contact cartridge fuses rated up to 60 amperes (a and e) and three knife-blade-contact cartridge fuses rated for current demands greater than 60 amperes (b, c, and d). Since the dimensions of a cartridge fuse depend on its current rating, it is not practical to substitute a high-current fuse for a low-current fuse. This is a safety design feature. In addition to the 250- and 600-volt ratings, a 300-volt (Class G) cartridge fuse is also available. Most cartridge fuses are manufactured in both conventional and time-lag designs. Some are designed for one-time service, and others have renewable fuse links. The advantage of the renewable type is its lower replacement cost.

Cartridge fuses are inserted into fuse clips rather than sockets. These clips match the contact design of the particular fuse. They are usually contained in a service box with a disconnect switch on the outside. This type of service panel is shown in Figure 4–19 and is generally called an enclosure with an E-X-O (externally operated) switch and a fuseholder. The disconnect switch is usually mounted above the fuseholder. An E-X-O switch is a safety switch that is rated for at least as much current as the fuse and is often rated for a higher current than the fuse. As noted earlier, the current rating of a fuse is equal to the current rating of the branch circuit that it protects.

To avoid a serious shock or electrocution, great care must be taken when a person is replacing cartridge fuses. First, the disconnect switch that controls the circuit *must be turned off*. It is also good practice to use a fuse puller, a nonconducting pair of grippers as pictured in Figure 1–19. After the blown fuse is replaced, but before the disconnect switch is turned back on, it is good practice to turn off all motors and equipment. This prevents possible damage to the equipment. Then start up the motors or other equipment units one-by-one. If a motor, for example, draws a fault current (greatly excessive current), the fuse will blow promptly. A fault can be caused by a motor defect or by a line short circuit. Before trying to troubleshoot a defective motor or other unit, turn off the disconnect switch at the fuse box. Make a visual inspection for evidence of charred insulation or melted copper. Check suspected short circuits with an ohmmeter. Correct the fault before the disconnect switch is turned on again. Since a cartridge fuse has the same external appearance whether it is blown or not, it is good practice to test

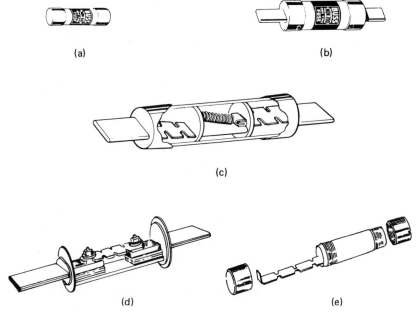

(a)

(b)

(c)

(d)

(e)

Figure 4-18 Cartridge fuses (a) ferrule-contact type (b) knife-blade contact type (c) girder-type, cover removable (d) cartridge-type ends removable

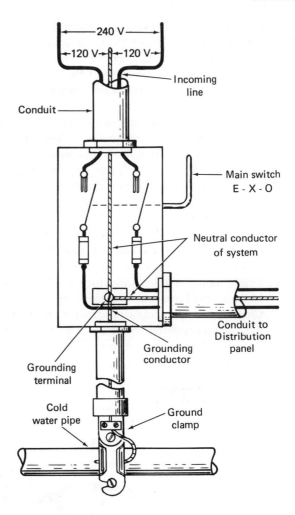

Figure 4-19 Arrangement of a service panel with cartridge fuses and an E-X-O Switch

the fuses right in the box before replacing them. Electricians often use a handy neon-light circuit tester, such as shown in Figure 1–19. This small device glows whenever more than 60 volts is applied to its terminals. If the test leads or probes are touched to the ends of a cartridge fuse and the neon light glows, the fuse is blown and there is a load on the circuit. If the fuse is not blown, the tester remains dark.* It is very poor practice and highly dangerous to test in live circuits unless electrician's insulating gloves are

---

*This test requires that at least a small load, such as an incandescent lamp, be switched into the circuit.

worn. A universal tester used for checking fuses, circuits, cords, and appliances is shown in Figure 1–19. It can be used to find current leakage in appliances and motors, and to check for defective cords and find short circuits. These two simple testers are lightweight and compact and are easier to use than an electrical meter. However, an ohmmeter must be used to check fuses in a "dead" circuit.

A *circuit breaker* is used on the large-load electric lines.

4–6 TYPES OF CIRCUIT BREAKERS

A circuit breaker looks like a toggle switch but differs in that it opens the circuit automatically when an excessive current demand hits the line. It acts like an oversize fuse to protect the line and the equipment on the line. Unlike a fuse, a circuit breaker can be reset to close the circuit, rather like the toggle switch. In case an overload still exists, the breaker opens promptly.

A basic circuit breaker is operated by a heater element and a bimetallic strip or by passing current through the bimetallic strip itself to generate the heat, and in turn to open the circuit. This is shown in Figure 4–20. The design that uses a separate heater element operates similarly. Basically, the breaker consists of a heat-sensitive bimetallic strip, two contacts, and a handle for resetting. As extra current flows through the metal strip, the heat produced causes the strip to bend, as depicted in Figure 4–20(b). With enough heat, the strip opens the circuit by springing apart and separating the contacts.

Figure 4–21 shows the handle positions for a circuit breaker. To reset the circuit breaker, the handle must first be moved to *reset* position, then to *on* position. A circuit breaker can be operated as a switch but it is not classified as a switch.

A circuit breaker in average use can carry its rated current load indefinitely. It will also carry a 50% overload for about a minute, a 100% overload for about 20 seconds, and a 200% overload for about 5 seconds. This is usually ample time for a motor to start without tripping the breaker. When nearly instant breaker action is required, a magnetic type of breaker is employed. It is easy to find a circuit breaker that has been tripped in a service panel because the handle will be resting in its off position.

A fuse box can be converted to a circuit breaker unit by using a type of breaker that has an Edison base. This type of breaker has a reset button and can be put back into operation simply by pushing this button. When this breaker trips, a color ring becomes visible to show that the circuit is open. That is, the button "pops up." It is advantageous to convert a fusebox to a breaker unit, because a breaker cannot "run out of fuses."

In another overcurrent device, a fuse and switch are used with a pilot

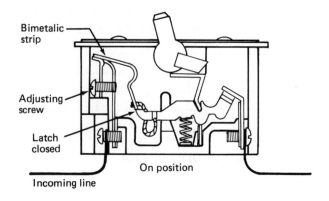

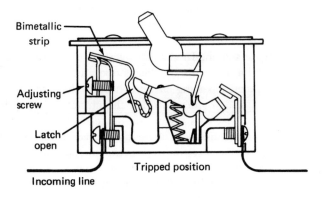

(a)

Latch

Bimetallic strip

To circuit

Fixed contact point

Movable contact point

Spring

To circuit

Fixed contact point

Movable contact point

(b)

Figure 4-20 (a) Construction of a thermal-type circuit breaker (b) action of bimetallic strip

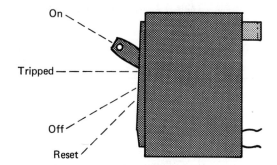

Figure 4-21 Handle positions for a circuit breaker

light, as shown in Figure 4–22. This combination is generally called a Switch-fuse. This has a plug-in holder that carries a Class G cartridge fuse. When the fuseholder is removed, the circuit is open. If the fuse is blown, the circuit is open. Note that the circuit can be opened by means of the switch if the fuse is not blown. The pilot light glows if the fuse is blown and the switch is closed. When the fuseholder is removed and the fuse is replaced, the pilot light does not glow when the fuseholder is inserted. However, if the fuse-holder is inserted upside down, the pilot light continues to glow, and the circuit remains open.

The advantage of the Switchfuse arrangement is that it almost eliminates the possibility of shock while the fuse is being replaced. Note that a different type of fuse cabinet is required for installing Switchfuses, compared with the cabinets used for conventional fuses or circuit breakers.

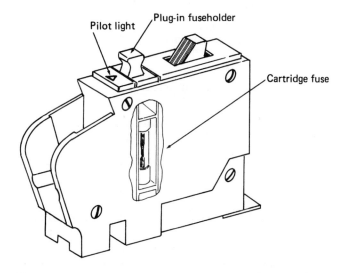

Figure 4-22 A Switchfuse overcurrent device

Table 4-1 Typical power consumption values for appliances

Appliance	Typical wattage
Air conditioner	1100
Attic fan	400
Blanket	200
Broiler	1000
Clock	2
Clothes dryer	4500
Coffee maker	1000
Deep fryer	1320
Dehumidifier	350
Dishwasher-disposer	1500
Egg cooker	600
Fluorescent lights (each tube)	15-40
Freezer	350
Fuel-fired heating plant mechanism	800
Griddle	1000
Hair dryer	100
Heat or sun lamp	300
Hot plate	1500
Iron, dry or steam	1000
Ironer	1650
Lamps, each bulb	40-100
Mixer	100
Oil burner	250
Portable fan	100
Portable heater	1650
Radio	100
Range, electric	8000
Refrigerator	200
Roaster	1380
Room cooler	600
Rotisserie	1380
Sandwich grill	1320
Toaster	1200
TV	350
Vacuum cleaner	300
Ventilating fan	400
Waffle iron	1320
Washer	700
Waste disposer	500
Water heater	2500
Water pump	700

When a replacement fuse blows as it is inserted, there is usually an overload somewhere on the line. To repeat a practical rule, before concluding that there is a fault in the wiring system, make sure that an excessive number of appliances are not plugged into a circuit. Table 4–1 gives a list of the electric current demands of home appliances, which is useful to keep for quick calculations. As a convenient example, let us assume that the line voltage in a house is 100 volts or 200 volts, as the circuit requirements may be. Then, a toaster that needs 1,200 watts of current would draw 12 amperes at 100 volts. Or, a range that consumes 8,000 watts would draw 40 amps at 200 volts.

Circuit breakers that are used in industrial installations are usually the magnetic type. Here, when the current exceeds a predetermined value, the mechanism is tripped by an electromagnet and the contacts are separated by spring action, as depicted in Figure 4–23(a). An industrial breaker is

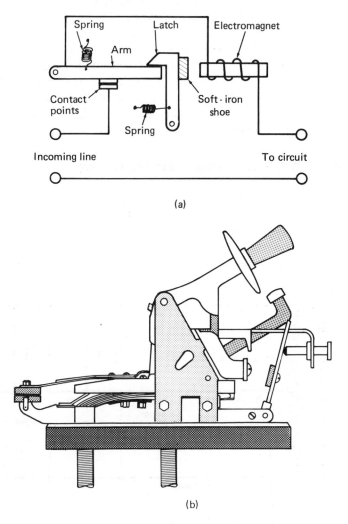

(a)

(b)

Figure 4-23 (a) Principle of the magnetic circuit-breaker (a) basic construction (b) typical appearance

shown in Figure 4–23(b). After this breaker has been tripped, it must be reset manually. Also, one essential feature for good breaker operation is very fast opening of the contacts in order to minimize arcing (electrical flame). Otherwise, the contacts will be burned or melted from the heat of arcing. This is the reason why a strong spring under tension is used to pull the contacts apart when the breaker is tripped.

The generating stations and substations that supply power to a large area use large and massive circuit breakers. These are similar to industrial breakers in their operation but are far more elaborate and complex so that they are able to control the very large current values that are involved. Very fast separation of contacts is essential here, and these breakers are designed to open a circuit within eight cycles (about 0.13 second) after the surge of a fault current. Arc suppression is also essential in these big circuit breakers, and various means are used to quench, cool, blow out, or divert the arc flame. Many of the larger breakers employ strong air blasts as the contacts are opening. Various other techniques are also utilized. In these heavy electrical load designs, the breaker contacts open at top speed under unusually heavy spring loading, and the breaker is tripped magnetically.

4–7 GROUNDING DEVICES AND METHODS

Each wiring system must include grounded circuits for the protection of personnel and equipment. Properly grounded circuits reduce shock and fire hazards greatly. Personnel must be protected against shock or electrocution if equipment becomes defective, and machinery must be protected against damage from lightning strikes, high voltage caused by crossed wires

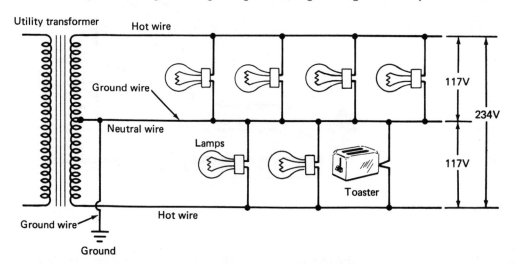

Figure 4-24 An example of a ground wire and a grounded neutral wire

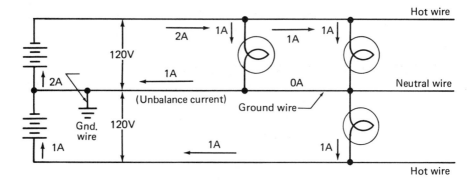

Figure 4-25 Example of a circuit with 1 ampere of unbalance current

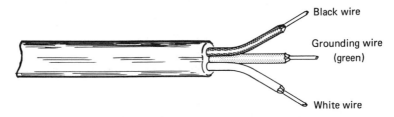

Figure 4-26 A ground wire in an NM (nonmetallic) sheathed cable

during wind storms, or transformer breakdowns in the utility system. All grounded circuits and grounding devices are listed in the National Electrical Code (NEC), which is discussed.

One example of a grounding system is shown in Figure 4–24. Here the neutral wire in a three-wire (hot/neutral/ground wire) system is connected to some ground (usually a cold-water pipe) by means of the ground wire. This connection between the neutral wire and ground is made at the panel at the service entrance. Electricians usually speak of the neutral wire as a *grounded wire*. In other words, that grounded neutral wire is different from the ground wire itself. Note that the grounded neutral wire carries any unbalanced current between the 120-volt circuits (Figure 4–25) but the ground wire has no current during normal operation. If the wiring system is struck by lightning, however, some lamps may burn out and other damage may occur, but that circuit from the grounded neutral wire to earth protects the wiring system from more extensive damage.

Next, observe the ground wire in the Type NM nonmetallic sheathed cable shown in Figure 4–26. This (green) grounding wire is connected to the metal parts of lighting fixtures, to metal housings of appliances, and to outlet boxes. One end of the ground wire is connected to earth ground. Its function is to protect people in case an appliance or device becomes defective. For instance, if an electrical leak takes place between a refrigerator motor and the metal case of the refrigerator, the ground wire ensures that a person who is standing on a damp kitchen floor will not get shocked when he touches the refrigerator door.

4–8 INSTALLATION OF GROUND WIRE

Figure 4–27(a) shows how a ground wire is installed from the neutral wire of the service switch to a water pipe. A 200-ampere service generally employs No. 4 gage ground wire which may be run in conduit or in armored cable. Rigid conduit looks like water pipe and comes in 10-ft. lengths. It

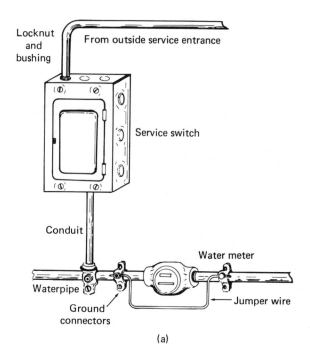

(a)

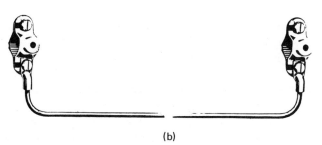

(b)

Figure 4-27 Ground connection hardware (a) arrangement for a ground wire (b) a meter shunt used for grounding a system

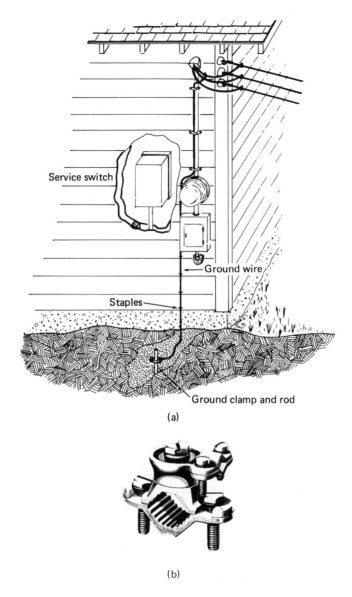

Service switch

Ground wire

Staples

Ground clamp and rod

(a)

(b)

Figure 4-28 (a) Ground wire arrangement in a rural installation (b) ground clamp for direct connection to a ground wire (Courtesy Buchanan Electric Products Inc.)

has a smooth inner surface that makes the insertion of wires easier. *Electrical metallic tubing (EMT)* has a rather thin wall and is a widely used type of conduit. Conduit is threaded like water pipe, but EMT is installed with special fittings. In Figure 4–27(a), note that the water meter must be jumped

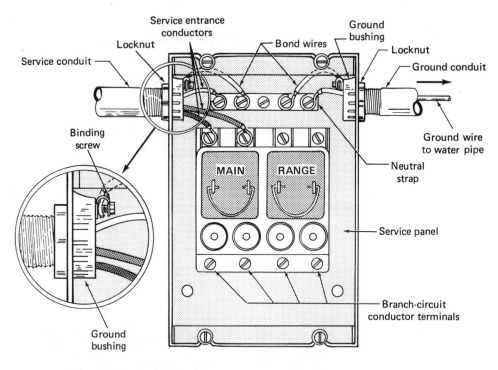

Figure 4-29 Grounding circuit at a service switch

with a *jumper wire*. This wire is often called a *meter shunt* and is shown in Figure 4–27(b). A typical ground clamp for use with conduit is shown in Figure 4–27(c).

In rural wiring systems, the ground wire is not connected at the neutral wire of the service switch. Instead, the ground wire is connected outside the building to the neutral wire coming in overhead from the main power pole and is brought down the outside of the house or down a yardpole (Fig. 4–28). Then the ground wire is connected to an underground water-pipe system or to a ground rod. An 8-ft length of galvanized-iron pipe 3/4 in. in diameter may be used as a ground rod, or an 8-ft length of copper rod 1/2 in. in diameter may be used. The ground rod must be located at least 2 feet away from the building and driven 1 foot below the surface of the soil. Figure 4–28(c) shows a ground clamp that is used to connect a ground wire.

Observe in Figure 4–29 that bond wires are connected from the service conduit and from the ground conduit to the neutral strap. The neutral strap is so-called because it is connected to the neutral wire at the service-entrance conductors. A bond wire is a short length of wire used to complete a ground circuit between metal parts. In this example, a bond wire completes the ground circuit between a ground bushing and a neutral strap. The ground wire connects the neutral strap to a cold-water pipe.

Note that the neutral strap is a metal bar with several terminals mounted in the service panel. The neutral strap is used as a common ground-connection point for the service-entrance conductors and for the branch-circuit conductors. Thus, the neutral strap is at ground potential—this means that there is no voltage difference between the neutral strap and earth ground (cold-water pipe).

Note that a ground bushing has a binding screw for connection of a bond wire, as shown in Figure 4–29. A ground bushing is a threaded metal fitting that secures the conduit to the wall of the service cabinet, so that the cabinet will be at ground potential. Observe in the figure that the ground bushing applies pressure against the inside wall of the cabinet, and that a locknut applies pressure in the opposite direction against the outside wall of the cabinet. A binding screw is a terminal screw (a machine screw) with a slotted hexagonal (six-sided) head used to make an electrical connection between two metallic parts—in this example, between the ground bushing and the bond wire.

4–9 INSTALLATION OF GROUND WIRES

At this point, it is helpful to briefly review some basic definitions. A ground is a conducting connection, whether intentional or accidental, between an electric circuit or equipment and earth, or to some conducting body which serves in place of the earth. A grounding receptacle is an outlet (socket) that has two parallel slots that mate with the two blades of a standard plug, plus a third U-shaped or round hole that mates with the third prong of a corresponding grounding plug. As explained in greater detail subsequently, this third prong is connected to a third or grounding wire in the cord that runs from the plug to the appliance (load). Grounding receptacles are used to avoid the danger of shock to the operator of an appliance in case the appliance might become defective. In other words, the use of a grounding receptacle places the frame of an appliance at ground potential. Therefore, the operator will not receive a shock from the frame in case the appliance becomes defective and a "hot" wire makes accidental contact with the frame.

As shown in Figure 4–26, the nonmetallic sheathed cable contains a green ground wire. This wire is connected to the green 6-sided hex screw on a grounding receptacle, shown in Figure 4–30(a). This green 6-sided hex screw connects to the mounting yoke of the grounding receptacle, shown in the figure, so that the outlet box is grounded. Note that the green hex screw also connects with the grounding terminal inside the receptacle box, shown in Figure 4–30(c). This is the means by which the housing of an appliance becomes grounded when it is plugged into a grounding receptacle. In Figure 4–30(b), note that the cord cap (plug) on the appliance cord has a U-

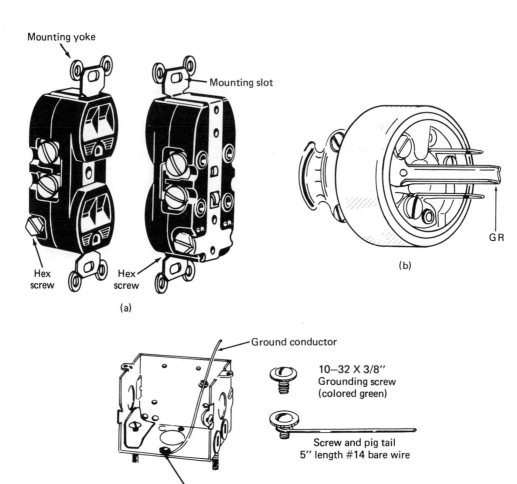

Figure 4-30 Grounded receptacle and plug (a) receptacle terminal arrangement (b) construction of cord plug with tandem blades (c) metal outlet box

shaped grounding blade marked (GR). The appliance cord connects the metal housing of the appliance, through this U-shaped GR blade, to the grounded receptacle. Thereby, the housing is automatically connected to the green ground wire in the cable that is connected to the outlet.

Figure 4–31 shows the arrangement of the grounding circuit. Observe that when a cord cap is plugged into the receptacle, the grounding blade of the cap makes contact with the green hex grounding screw, and thence to the green ground wire in the sheathed cable (or conduit) that runs from the receptacle to ground (a cold-water pipe). As explained previously, the

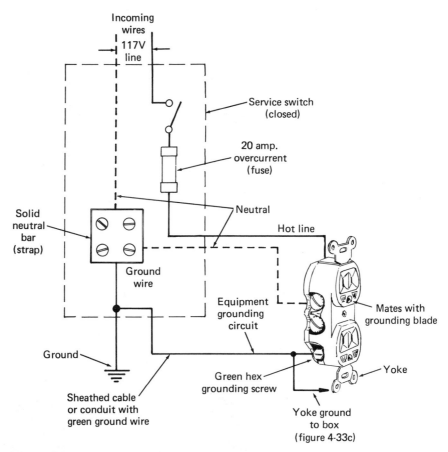

Figure 4-31 Arrangement of a grounding system

green hex grounding screw also connects to the yoke of the receptacle. Of course, the neutral wire is also grounded, as was explained before. In summary, the advantage of using a grounding receptacle is protection against accidental shock to the operator. Since the frame of the appliance is connected by the grounding blade of the plug to the green ground wire in the sheathed cable, the frame remains at ground potential even though there might be accidental electrical leakage from the "hot" side of the line to the frame of the appliance.

Some other grounded receptacles are shown in Figure 4–32. A 120/240-volt receptacle has parallel slots for the 120-volt circuit and tandem slots for the 240-volt circuit, as seen in (a). Note in (b) that a single receptacle with tandem blades may be used; it is rated for 240-volt circuits up to 15 amperes and is designed for small air conditioners. Large air conditioners also use a single receptacle but one with more rugged construction and rated

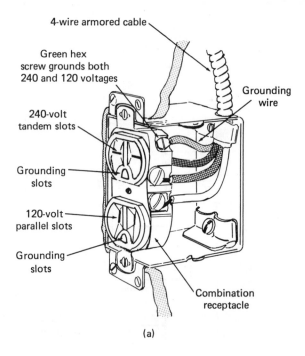

4-wire armored cable

Green hex
screw grounds both
240 and 120 voltages

240-volt
tandem slots

Grounding
slots

120-volt
parallel slots

Grounding
slots

Grounding
wire

Combination
receptacle

(a)

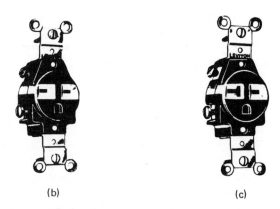

(b)   (c)

Figure 4-32 A 120/240 volt receptacle (b) single
receptacle with tandem blades (c) single receptacle with
a horizontal and a vertical blade

for operation in the higher 21- to 30-ampere range. Another type of single
receptacle with a horizontal and vertical blade is shown in (c). This is rated
for 240-volt operation at maximum current of 20 amps. It is used with air
conditioners, power tools, or garden equipment.

Figure 4–33(a) depicts a surface-type receptacle with an L-shaped ground
slot. It is rated for 120 or 240 volts and for currents up to 30 amperes. This

Figure 4-33 A surface-type receptacle with an L-shaped ground slot (Courtesy Leviton Inc.)

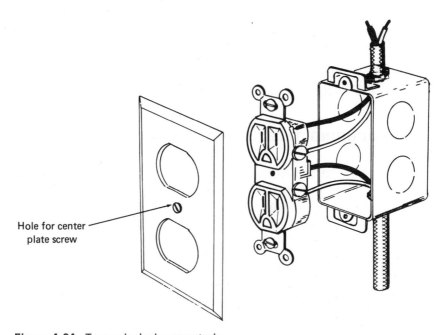

Hole for center plate screw

Figure 4-34 Two-pole duplex receptacle

type of receptacle is self-contained and is not installed in an outlet box, but is mounted right on a wall. It is often used for a clothes dryer. Another surface-type receptacle, similar to this, does not have the L-shaped ground slot but is rated for operation at either 120 or 240 volts and currents up to 50 amperes. This type of receptacle is often used with high-speed dryers or

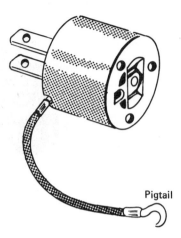

Figure 4-35 Adapter for use of 3-pole cord cap with a two-pole receptacle

electric ranges. Note that three-pole receptacles were preceded by two-pole (two-slot) receptacles (Figure 4–34), some of which are still in use. A two-pole receptacle will serve a modern three-pole cap if a special adapter is used, as shown in Figure 4–35. Note that the pigtail (extra wire and hook) of the adapter should be attached to the center screw of the outlet box.

4–10 GROUNDING TELEVISION ANTENNAS

A roof-mounted television antenna is a form of lightning rod and must be properly grounded to avoid lightning hazard. Grounding the mast is the chief requirement. This is shown in Figure 4–36.

A bare copper wire is connected to the mast, then run as directly as possible to a cold-water pipe or a ground pipe. This copper wire is stapled at intervals to the outer wall of the building. Either a ground strap or a ground clamp, may be used.

Since lightning could strike the antenna lead into the house as well as the mast, a *lightning arrestor* is attached to the lead-in near the point of entry into the building. A bare copper ground wire is attached to the arrestor and run to a cold-water pipe or ground rod.

Most TV antenna installations use twin leads, although TV coaxial cable is becoming more widely used. The lightning arrestor provides a short spark-gap arrangement to ground. A spark gap consists of a pair of metal points separated by a short air gap. Thus, the arrestor is normally an open circuit. However, if a lightning or other high-voltage surge travels down the lead-in, the air molecules in the gap ionize and provide a low-impedance

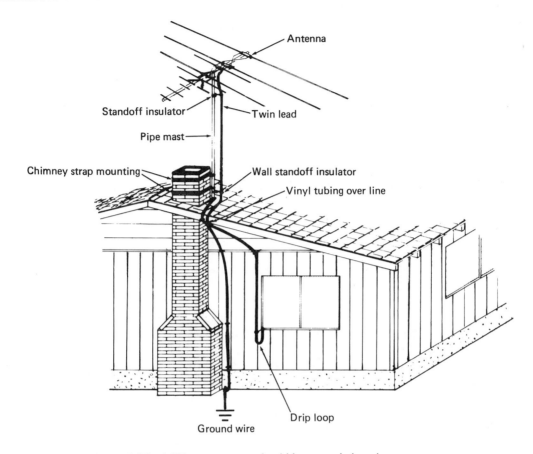

Figure 4-36 A TV antenna mast should be grounded as shown

path to ground. Thus, nearly all of the surge is shunted to ground at the arrestor and does not enter the building.

4–11 GROUNDING AUTOMOBILE ELECTRICAL SYSTEMS

A ground in the wiring system of a building is a connection to earth. A ground in an automotive electrical system is not a connection to earth; it connects to the metal frame of the car. An automobile ground is actually insulated from earth by the rubber tires of the car. It does not serve a protective purpose and is used simply for reasons of economy.

A vehicle ground is not necessarily at ground potential. Therefore trucks that transport flammable cargoes such as gasoline and liquid propane gas often employ some means of discharging static electricity. For example, a chain may be suspended from the chassis so that it touches the pavement.

This ensures that there will be no sparking of static electricity, which could cause an explosion.

Some airplanes use sharp point-dischargers for dissipating static electricity charges into the air, since static electricity leaks off easily from a needle point.

A vehicle ground may be either a positive ground or a negative ground. If the negative terminal of any vehicle battery is connected to the metal structure of the vehicle, a negative ground system is employed.

4–12 GROUND FAULT CIRCUIT INTERRUPTER

A special type of circuit breaker is installed in the outdoor outlet boxes of modern residential and commercial wiring systems. Lawn, patio, and swimming-pool outlets are thus provided. This type of breaker is called a *ground fault circuit interrupter*. It is abbreviated GFCI or GFI. The GFCI is generally mounted on the outlet box. The cord of the appliance is plugged into this interrupter, so the current flows through the interrupter, which acts as a protective device, then on to the appliance. As will be explained later, a ground fault denotes a defect in the insulation of an appliance. Such a defect makes the case or frame of the appliance "hot" and imposes a shock hazard. However, the GFCI quickly switches the circuit open if a fault occurs and the appliance case gets "hot." In other words, the GFCI cuts a power source off the line. This interrupter action makes the appliance inoperable and protects the operator from the possibility of shock.

Note that a GFCI does not act like a circuit breaker or a fuse. A circuit breaker or fuse opens a circuit only when the current in the line is excessive. When the frame of an appliance becomes "hot," the line current demand is not necessarily high but a person can be killed by a current flow of 1/15 ampere. The typical circuit breaker or fuse does not open until the current exceeds 15 or 20 amperes. Acting as a protective device, the GFCI opens the circuit to the appliance whenever ground-fault current exceeds 0.005 ampere. Note too that a GFCI acts very quickly and opens the circuit in about 0.125 second.

STUDENT EXERCISE 4-A

This exercise provides familiarity with basic electrical hardware and supplements the illustrations in this chapter.

1.   Obtain a standard metal outlet box, an SPDT toggle switch, a duplex receptacle outlet, a wall plate for a toggle switch, and a wall plate for a duplex outlet. Necessary machine screws are supplied with the hardware.

2.   Observe the provisions that have been made for mounting the box on a stud with nails.

3. Note the scale which may be stamped on the side of the box to indicate the distance that the box extends beyond the stud. This scale makes it easy for the electrician to allow for various thicknesses of wallboard, panelboard, etc.

4. Attach the toggle switch to the outlet box with the machine screws provided. Note how the slots in the switch strip permit the switch to be shifted a little one way or the other.

5. Place the plate over the toggle switch and attach to the outlet box with the machine screws provided. Note the relations of the parts and the snug fit of the wall plate.

6. Remove the wall plate and toggle switch from the outlet box. Attach the duplex receptacle to the outlet box with the machine screws. Note the mounting slots in the receptacle strip. (See Figure 4-30.)

7. Attach the wall plate to the outlet box, over the duplex receptacle with the machine screw that is provided. Note the relations of the parts and the snug fit of the wall plate.

8. Suggest a reason why it is desirable to have mounting slots in switch and outlet strips. (See Figure 4-1.)

STUDENT EXERCISE 4-B

This exercise provides practical experience in ganging metal boxes.

1. Obtain a pair of metal outlet boxes and single and double wall plates.
2. Observe the construction of the metal boxes.
3. Remove the wall end of each box, as shown in Figure 4-37.
4. Fit the open sides of the boxes together.
5. Tighten the screws and secure the two boxes together as one unit.
6. Does this ganged outlet box require a pair of single wall plates, or a double-sized wall plate?
7. Could you install a toggle switch and a duplex receptacle in the ganged box?
8. Could you install a pair of duplex receptacles in the ganged outlet box?

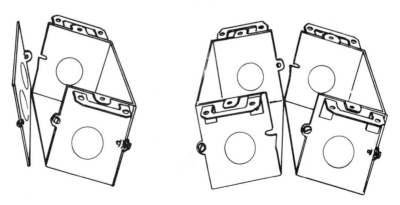

Figure 4-37 Ganging metal boxes

9.   Is it possible to gang three or four metal boxes?

10. List several combinations of devices that could be installed in a three-gang outlet box.

STUDENT EXERCISE 4-C

This exercise provides familiarity and practical experience with basic ground clamps and straps.

1.   Obtain the two types of conduit ground clamps and a ground strap, as shown in Figure 4-38. Also obtain a short section of ordinary water pipe and a short length of No. 12 bare copper wire.

2.   Inspect the conduit ground clamps and observe the provisions for securing each clamp tightly around a water pipe. Also observe the terminals for connecting a ground wire to the clamp.

3.   Slide the ground clamps over the water pipe and tighten them securely in place.

4.   Insert a copper wire through the threaded conduit opening in each clamp and connect the end of the wire to the ground clamp.

5.   Next, inspect the ground strap and see how it is tightened in place after being wrapped around the water pipe.

6.   Mount the ground strap on the water pipe and tighten it securely.

7.   Connect the end of a copper wire to the ground strap and tighten the terminal nut.

8.   Approximately what diameters of water pipe does each of the two ground clamps accommodate?

9.   What diameter of conduit does each of the ground clamps accommodate?

10. Approximately what diameters of water pipe does the ground strap accommodate?

11. Would it be good practice to scrape or sandpaper the surface of the water pipe that is contacted by the ground clamp or strap?

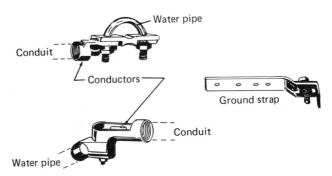

Figure 4-38 Standard ground clamps for conduit and a ground strap.

QUESTIONS

1. What is the purpose of using a grounded circuit in a wiring system?
2. What is the difference between a ground wire and a grounded neutral wire?
3. What is the purpose of a ground wire being connected to a refrigerator?
4. What is rigid conduit and how does it differ from electrical metallic tubing?
5. What is a ground rod and how is it used?
6. Why is the neutral strap given that name?
7. Why must a roof-mounted television antenna be grounded?
8. How does a ground fault circuit interrupter differ from a circuit breaker?

5 CONNECTIONS AND SPLICES

5-1 GENERAL CONSIDERATIONS

A splice consists of two wires joined together electrically and mechanically at their ends. A tap consists of a wire that is connected at one end to another continuous wire. Splices and taps must be properly made, or trouble conditions will occur. A poor connection has abnormal resistance and a corresponding abnormal I^2R loss. Excessive heat is produced at the defective connection, and an excessive voltage drop occurs across the connection. Insulation may catch on fire, or the wires may melt. To ensure good electrical connections, it is necessary to solder all splices and taps, unless approved solderless connectors are used. It is essential that wires be properly stripped or "skinned" at connection points, so that contacting surfaces are bright and clean.

Insulation way be removed from wires at connection points by means of a jack knife. It is important not to score or nick the conductor with the knife. The insulation should be removed by cutting at an angle, as in sharpening a pencil with a knife. Figure 5-1(a) shows the right and wrong ways to remove insulation with a knife. About three inches of wire is usually exposed. If a tin coating is present, it should not be scraped off because the tin helps to make a good solder connection. To splice a pair of wires together, cross the wires about 1 inch from the insulation, as shown in Figure 5-1(b). Then, make six to eight turns as shown, using your fingers to start and a pair of pliers to finish the job.

A soldering iron, gun, or torch is used to solder the joint. Figure 5-1(c) shows how a soldering iron is used. Electric soldering paste is first applied to the wires. Then, the wires are heated with the soldering iron. Solder is applied to the wires on the opposite side from the iron and allowed to flow

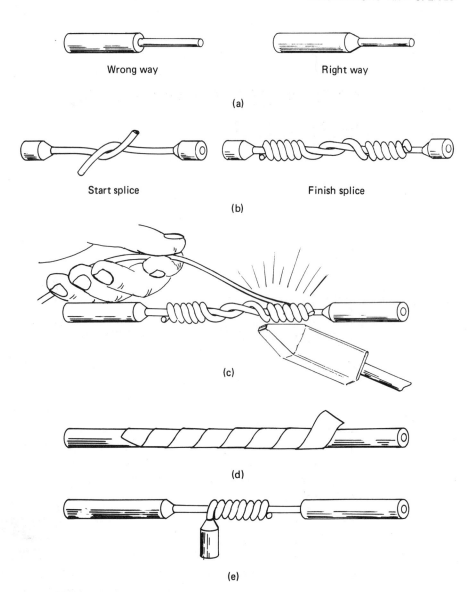

Figure 5-1 (a) Right and wrong ways to remove insulation with a knife (b) wires are twisted together to make a splice (c) proper use of a soldering iron (d) spiral wrap with plastic tape (e) appearance of a tap splice

smoothly throughout the connection or joint. A soldered joint may be covered with electricians' rubber tape, followed by a layer of friction tape. Plastic tape is often preferred, however, because it is easier to apply. The joint is completely covered with a spiral wrap, as shown in Figure 5–1(d). A properly taped connection is waterproof. Electricians' plastic tape is also acid-proof.

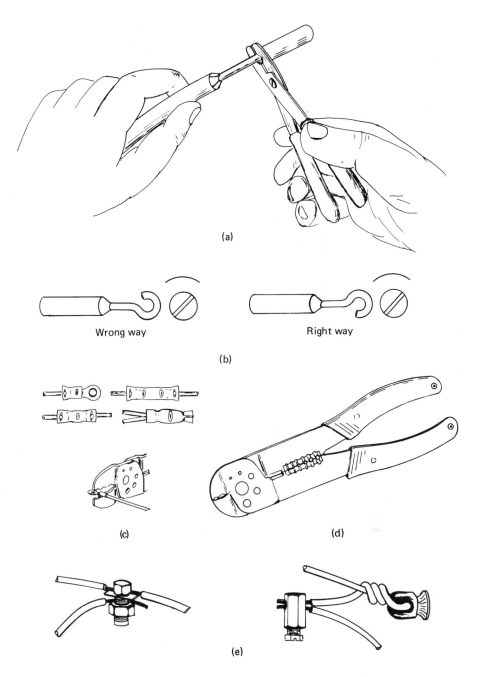

(a)

Wrong way

Right way

(b)

(c)

(d)

(e)

Figure 5-2 (a) Combination wire cutter and stripper (b) loop connection at a screw terminal (c) examples of solderless terminals and connectors (d) multi-purpose tool used to crimp solderless connecting devices (e) two other types of solderless connectors

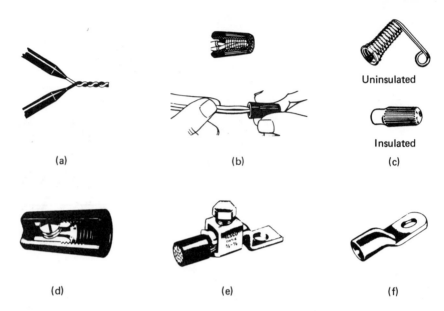

Uninsulated

Insulated

(a)   (b)   (c)

(d)   (e)   (f)

Figure 5-3 (a) A pigtail splice (b) wire nut (c) spring-type connectors (d) screw-type solderless connectors (e) screw-type solderless lug (f) solder-type lug

Figure 5–1(e) shows how a tap splice is made. It is soldered and taped in the same manner as an end splice. Tap (or tee) splices are used only if the tap wire does not put or impose a substantial strain on the continuous wire. Note that soldering paste should be noncorrosive, such as the resin-base type. Electricians' solder is 50% tin and 50% lead, or 50/50 grade.

Electricians often use a combination wire cutter and stripper, as shown in Figure 5–2(a). This tool cuts and strips clean either solid or stranded wire. It may also be used to bend loops at the ends of wires for connection under screw terminals, as in Figure 5–2(b). Note that the loop should have the same direction as the screw turns while being tightened. Solderless terminals, lugs, and connectors are often used, as shown in Figure 5–2(c). To make a solderless splice or to secure a solderless terminal in place, indenting pressure is applied with a multipurpose tool, as shown in Figure 5–2(d). Two other types of solderless connectors are shown in Figure 5–2(e). Type A, at left, is used to tap an existing line when there is more or less strain on the wires. Type B, at right, is used in service-entrance connections. Both types should be insulated with a winding of plastic tape.

Figure 5–3(a) shows a pigtail splice, which is approved provided there is no strain on the wires. Wire nuts or wing nuts are widely used to make solderless pigtail splices, as shown in Figure 5–3(b). A wire nut is seen in the illustration; a wing nut is quite similar, except that it has two "wings" on its outer surface for turning the solderless connector on the pigtail splice. Note that the splice is always twisted in a clockwise direction. Solderless connectors of most types are usually permitted for splicing copper wires, aluminum wires, or for splicing a copper wire to an aluminum wire. However, there are exceptions, and the electrician should check with local electrical codes, and read the information on the carton in which the solderless connectors are shipped.

Another type of solderless connector, called the spring-type connector, is pictured in Figure 5–3(c). It is similar to a wire nut, but a pigtail splice is not required. Instead, the ends of the wires are merely placed beside each other, and the spring-type connector is turned up tightly on the wires. Then, the lever end of the connector is broken off and the joint is insulated with plastic tape. If the insulated variety of spring-type connector is used, the procedure is the same as for a wire nut. A screw-type solderless connector is shown in Figure 5–3(d). A screw-type solderless lug is pictured in Fig. 5–3(e) and a solder-type lug is seen in Figure 5–3(f).

Figure 5–4(a) shows a screw-type fixture connector. Examples of various solderless crimp-type wire terminals are illustrated in Figure 5–4(b). Another variety of screw-pressure splicing device provided on some switches is shown in Figure 5–4(c). Solderless crimp-type connectors are available in the form seen in Figure 5–4(d). Note that some electrical devices do not have exterior terminals. Instead, holes are provided into which the ends of wires can be pushed. An interior mechanism seizes the wire and provides a good connection. If the wire needs to be released at any time, a slot is provided

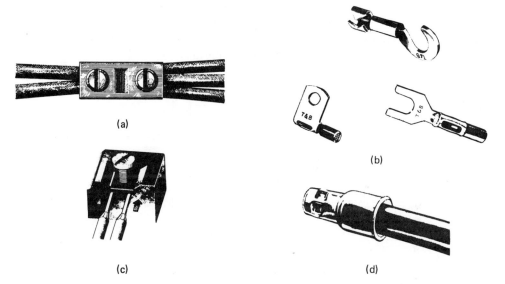

(a)

(b)

(c)

(d)

Figure 5-4 (a) A screw-type fixture connector (b) varieties of crimp-type wire terminals (c) a screw-pressure splicing device found on switches (d) another form of solderless crimp-type connector (Courtesy Buchanan Electric Products Inc.)

through which a screwdriver can be inserted. This is called a back-wired device.

Note that a "strip gage" is provided on the back of a duplex plug that uses slip-in wire connections. This is used by the electrician to check the length of the stripped end of the wire. If the exposed end is too short, it will not make proper connection; if too long, more or less bare wire will remain outside the device and constitute a hazard. Another type of back-wired device does not automatically seize the end of the wire when it is inserted. Instead, pressure screws are provided and must be tightened to secure the wire in the hole.

5–2 SPLICING STRANDED WIRE

Since lighting fixtures usually have stranded wire, it becomes necessary to splice the stranded wire to the solid building wire. The approved method is shown in Figure 5–5. The solid wire is bent back over the splice to provide greater mechanical strength. A joint of this type is usually soldered and taped, although wire nuts may be approved in some localities. Stranded copper wires are spliced as shown in Figure 5–6. The ends of the wires are fanned out, and then meshed as uniformly as possible. Several of the strands are wrapped around the conductor as shown, to provide mechanical strength. The left-hand strands are wound in an opposite direction from the right-hand strands. Finally, the joint is soldered and taped.

Tap or tee splices for stranded wire are made as shown in Figure 5–7. The continuous wire is separated into two halves and the tap wire is passed through. The tap wire is divided into two halves, and each is wound around

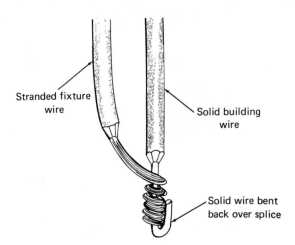

Figure 5-5   Splicing stranded wire to solid wire

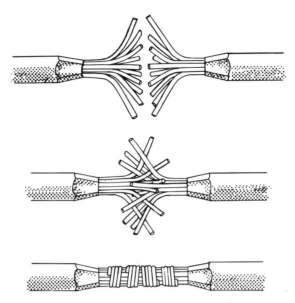

Figure 5-6 Splicing the ends of stranded wires

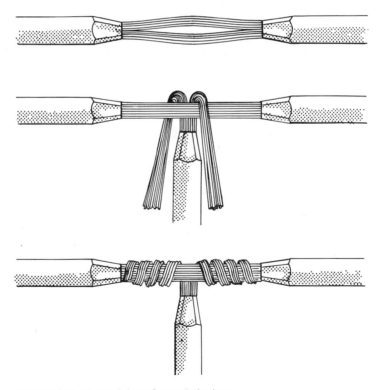

Figure 5-7 Tee splicing of stranded wires

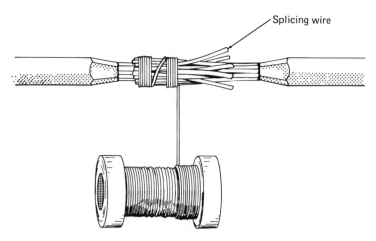

Splicing wire

Figure 5-8 Mousing a splice

the continuous wire in opposite directions. Finally, the joint is soldered and taped. When large stranded cable is to be spliced, the method shown in Figure 5–8 is employed. Note that the cable strands remain straight and are merely intermeshed. The strands are pounded or squeezed together with a wooden club or with a suitable pair of pliers. The joint is then wrapped with a separate copper wire of smaller gage. A small pigtail splice is made at the end to secure the wrap. This is called "mousing a splice," and No. 14 or 16 mousing wire is generally utilized. The completed joint is heated with a torch and thoroughly soldered.

5–3 MILITARY WIRE SPLICING

Somewhat stronger splices are required by the military services. One of the basic splices for solid wires is the Western Union splice shown in Figure 5–9. Approximately eight inches of end is utilized. The wires are twisted together for 1-1/2 inches, or three turns, and the ends are then bent at right angles to the direction of the joint. Each end is then close-wrapped around the wire for five or six turns. Finally, the ends are clipped off close to the joint. A tap or tee splice for solid wires is made much the same as a commercial splice, with at least eight turns, as depicted in Figure 5–10. If the tap wire is stranded, a combination seizing-wire splice is utilized, as seen in Figure 5–11. About 1 inch of insulation is removed from the end of the stranded wire, and the joint is then wrapped with the seizing wire. A seizing wire binds other wires together. Note that four turns are made ahead of the wrapping for added strength.

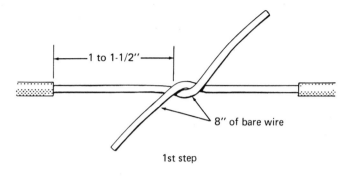

1st step

2nd step

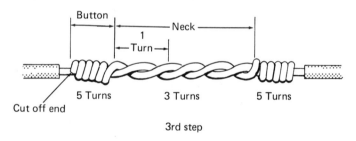

3rd step

Figure 5-9 The Western Union splice

Stranded wires are spliced by seizing with a length of solid wire, as shown in Figure 5–12. The stranded wires are first tied together with a square knot, and the seizing wire is placed through the knot. Then the seizing wire is wrapped around the ends of the stranded wires and up on the insulation. Finally, the leftover ends of the stranded wires and seizing wire are cut close to the joint. After wires have been spliced and the joint soldered, plastic tape is wrapped around the splice as shown in Figure 5–13. Note that the crotch laps (where the wires form a "Y") are taped with one close turn and two additional turns around the wire. Two layers of friction tape are then wrapped over the plastic tape, starting at the center and wrapping

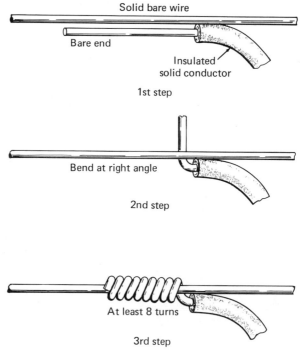

Solid bare wire

Bare end

Insulated
solid conductor

1st step

Bend at right angle

2nd step

At least 8 turns

3rd step

Figure 5-10   Tap splice for solid wires

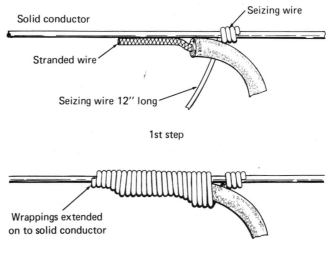

Solid conductor

Seizing wire

Stranded wire

Seizing wire 12" long

1st step

Wrappings extended
on to solid conductor

Figure 5-11   Tap splice for stranded wire to solid wire

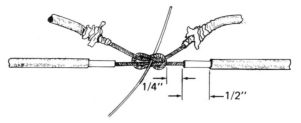

Seizing wire inserted through knot

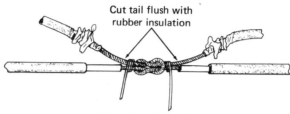

Cut tail flush with
rubber insulation

Wrapping seizing wire

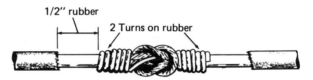

1/2" rubber

2 Turns on rubber

Splice of conductor after seizing is completed

Figure 5-12   Splicing stranded wires

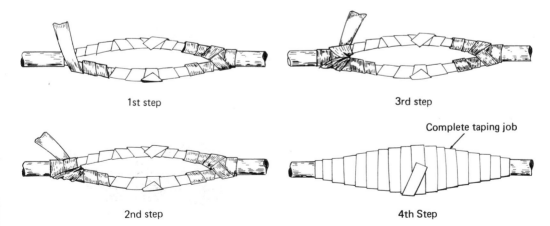

1st step

3rd step

2nd step

4th Step

Complete taping job

Figure 5-13   Taping a splice

to one end. The braid is covered for 1 inch, and the tap is then cut. The wrap is continued by covering the other braid for 1 inch and wrapping back, finishing the wrap at the center of the splice.

5–4 ESSENTIALS OF GOOD SOLDERING

Properly soldered joints require clean metal surfaces. Scraping is the usual method, using a knife. This is an easy job with solid wire, but requires time and care with stranded wire. Unless all the strands are clean and bright, it is difficult or impossible to make a good soldered joint. As noted previously, resin (or rosin) flux is required to make the solder flow properly. Acid flux should be avoided because it is destructive and defeats the purpose of soldering. Acid flux corrodes and weakens a wire. A soldering iron (copper) should be kept well tinned to avoid rapid oxidation and pitting. Small joints can be soldered with a soldering gun, for example, whereas large joints require a torch or a large soldering iron. Unless ample heat can be applied, the result will be a cold-soldered joint that may look all right, but will be defective both electrically and mechanically.

The best method is always to melt the solder against the wire—never against the soldering iron. If the joint is sufficiently hot to flow the solder, the possibility of a cold-soldered joint is avoided. However, an overheated joint is very undesirable because it will burn the insulation back along the wire. Note that, when soldering large stranded wires, it may be necessary to wipe the solder around the surfaces so that it will enter all the crevices and fill the inside of the splice evenly. Wiping can be done with a dry cotton rag.

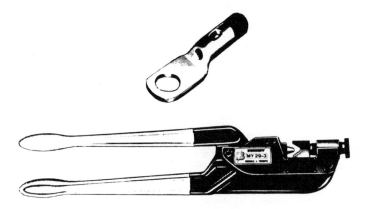

Figure 5-14 A heavy duty solderless lug and indenting tool (Courtesy Buchanan Electric Products Inc.)

Another problem involving large wires in tubular lugs is to obtain good penetration of solder down to the bottom of the lug. Since considerable experience is required for this task, large solder-type tubular lugs are prohibited in certain applications, and heavy-duty solderless lugs must be used instead. Figure 5–14 shows a typical lug and the indenting tool that is used to make a pressure connection.

STUDENT EXERCISE

This exercise provides practical experience in making simple connections.

1.   Obtain a standard junction box with cable clamps, a standard switch box with cable clamps, an SPDT toggle switch, five short sections of 14-2 NM cable, and three wire nuts.

2.   Strip the insulation from the wires at one end of each cable, leaving 3/4 inch of wire bare.

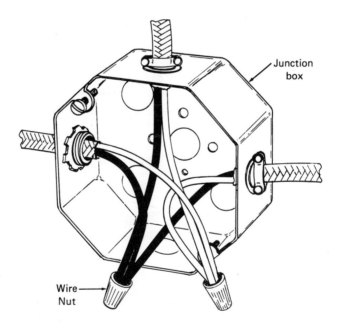

Figure 5-15 Wires connected by wire nuts

3.   Insert three of the cables into the junction box, as shown in Figure 5-15, and tighten the cable clamps.

4.   Connect (twist) the black wires together and connect the white wires together.

5.   Place a wire nut over each of the connections and turn it up tightly.

6.   Insert the remaining two cables into the switch box, as shown in Figure 5-16, and tighten the cable clamps.

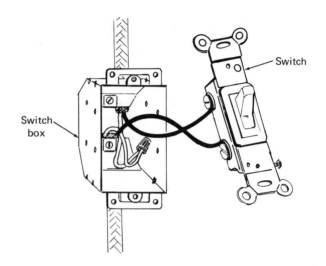

Figure 5-16 Wires connected to switch terminals

1. Cut away damaged wire

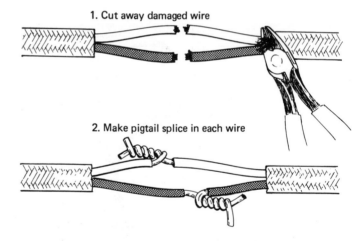

2. Make pigtail splice in each wire

3. Wrap each wire with tape

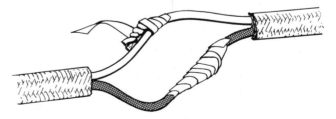

Three-step repair of damaged
cord using pigtail splices

Figure 5-17 Making pigtail splices

7.   Connect the white wires together, using a wire nut.

8.   Connect the black wires to the switch terminals. Make sure that the loops of the black wires are made in the same direction the terminal screws tighten.

STUDENT EXERCISE

This exercise provides practical experience in making pigtail splices.

9.   Obtain a damaged section of two-wire cable (an undamaged section could also be used), a pair of diagonal cutters or scissors, pliers, and plastic tape.

10. As shown in Figure 5-17, remove the outer insulation for about 2 inches from the damaged point in the cable.

11. Cut the wires at the damaged point and strip about an inch of insulation from the end of each wire.

12. Make pigtail splices and tighten with a twist of the pliers, if necessary.

13. Wrap each exposed pigtail splice with plastic tape.

STUDENT EXERCISE

This exercise provides practice in making an underwriter's knot and connecting the ends of a lamp cord to the terminals of a receptacle outlet plug.

14. Obtain a section of lamp cord and a standard plug for a receptacle outlet.

15. Remove the outer braid from the end of the cord, for a distance of approximately 1-1/2 inches.

16. Insert cord through plug.

17. Tie a knot in the pair of wires, as shown in Figure 5-18, and tighten the knot at the end of the braid.

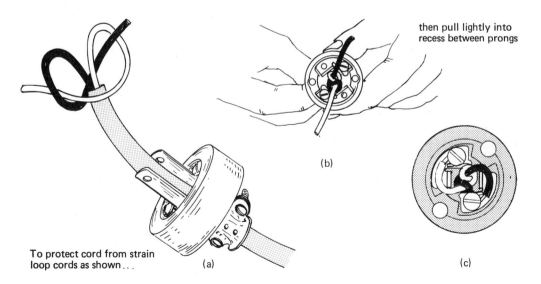

then pull lightly into recess between prongs

(b)

To protect cord from strain loop cords as shown...   (a)   (c)

Figure 5-18   Attaching a cord to a receptacle outlet plug

18.   Strip the insulation back from the ends of the wires for about 1/2 inch.

19.   Pull the cord so that the knot goes down into the recess between the prongs of the plug.

20.   Connect the bare ends of the wires under the terminal screws of the prongs.

QUESTIONS

1.   What is a splice?

2.   What are the requirements for making a good splice?

3.   What is the safest method of connecting a splice?

4.   How should the insulation be removed from a single-stranded wire?

5.   What are the advantages of using plastic tape instead of rubber tape or friction tape?

6.   What type of soldering flux should you use in splicing wires?

7.   Why should the loop in a wire be wrapped in the same direction as the screw tightens?

8.   What is the purpose of a wire nut?

9.   How can the electrician determine if it is permissible to use wire nuts to connect a copper wire to an aluminum wire?

10.   What is the description of a device in which the end of a wire is inserted and an interior mechanism seizes the wire?

11.   How is a wire released from a back-wired device?

12.   What is a strip gage?

13.   Why is a strip gage necessary for stripping a wire that is being inserted into a back-wired device?

14.   What is a tap or tee splice?

15.   What is the name of the method used for splicing large wires with a smaller wire?

16.   Why is a Western Union splice used?

17.   What is a basic requirement for a good soldering connection?

18.   Why is the solder always melted against the wire?

19.   Why are large soldering lugs prohibited in some applications?

REVIEW QUESTIONS FOR PART 1

1.   With reference to Figure I-1, battery B has a voltage of 12 volts, R1 has a resistance of 12 ohms, and R2 has a resistance of 6 ohms. How many amperes of current flow through R1? Through R2? What is the total amount of current in the circuit?

2.   How much power does R1 consume in Figure I-1? How much power does R2 consume? What is the total power consumption of the circuit?

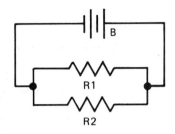

Figure I-1 Basic parallel circuit

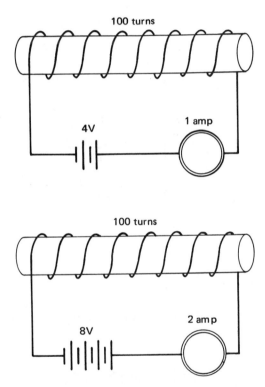

Figure I-2 Electromagnets with different currents

3. Two electromagnets are depicted in Figure I-2. Both magnets are the same, except that one is connected to a 4-volt battery and the other is connected to an 8-volt battery. In turn, one magnet draws 1 ampere and the other magnet draws 2 amperes. Which magnet is the stronger?

4. What is the resistance of the coil winding in Figure I-2?

5. How much power does each magnet consume in Figure I-2?

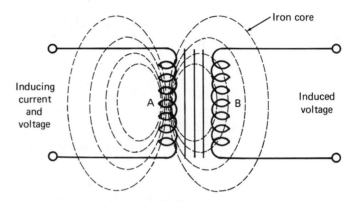

Figure I-3 A pair of magnetically coupled coils

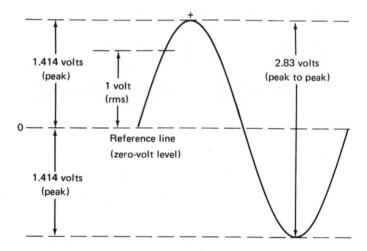

Figure I-4 Voltage relations in a sine wave

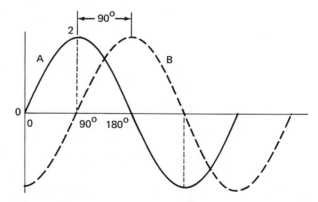

Figure I-5 Out-of-phase sine waves

6. What electrical device do coils A and B represent in Figure I-3? What name do electricians give to coil A? To coil B?

7. With reference to Figure I-4, what voltage would an electrician assign to the sine wave? If this waveform were applied to an ordinary AC voltmeter, what would the scale reading be?

8. What is the phase difference between the two sine waves depicted in Figure I-5?

9. How many kilowatt hours are indicated by the meter dials in Figure I-6?

10. With reference to Figure I-7, what type of conductor is the No. 4 gage wire? The No. 10 gage wire?

11. What kinds of insulation are commonly used on copper wires?

12. State an advantage and a disadvantage of aluminum wire in comparison to copper.

13. What are the names of the three types of conduit depicted in Figure I-8?

14. What is an outlet?

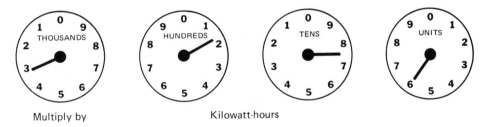

Multiply by   Kilowatt-hours

Figure I-6   Example of kwh meter reading

Actual Wire Size Of Copper Conductors

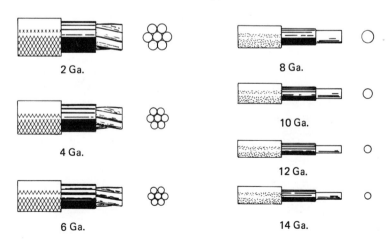

2 Ga.   8 Ga.

4 Ga.   10 Ga.

6 Ga.   12 Ga.

             14 Ga.

Figure I-7   Sizes of copper conductors

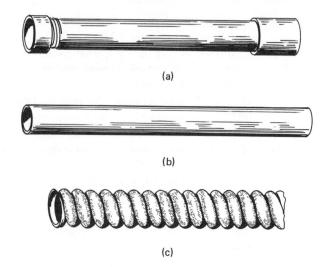

(a)

(b)

(c)

Figure I-8 Three basic types of conduit

15. How does a single-pole double-throw switch differ from a three-way switch?

16. How does a four-way switch differ from a three-way switch?

17. What is the basic difference between a switch and a circuit breaker?

18. How many fuses are installed in a 120-volt circuit? In a 240-volt circuit?

19. How does a time-lag fuse differ from a conventional fuse?

20. What is the difference between a ferrule and a knife-blade type of cartridge fuse?

21. How can a fuse be tested with a neon light?

22. To what is the neutral strap in a service panel connected? What is the basic function of a ground connection?

23. How does a lightning arrestor operate?

24. What is a GFCI?

25. How are soldered joints protected?

26. What is a solderless connector?

27. How are stranded wires spliced?

28. What is a Western Union spice?

29. Is seizing the same operation as stripping?

30. What is meant by "mousing" a joint?

WIRING LAYOUTS and APPLICATIONS

6 HOW TO MAKE HOME WIRING LAYOUTS

6–1 GENERAL CONSIDERATIONS

Adequacy is the first consideration in planning a wiring project or a wiring system. An adequate wiring system is efficient and is convenient for all intended uses. Note that this may involve planning for the future as well as present needs. Since rewiring is comparatively expensive, it should be avoided whenever possible. An adequate wiring system ensures that toasters and irons will heat quickly and that a minimum amount of electricity will be wasted as I^2R losses in the wires. An ample number of outlets at functional locations should be provided to avoid the nuisance of long and snarled extension cords. Moreover, an adequate wiring system minimizes fire hazard due to circuit overloading.

Figure 6-1 depicts the lighting provisions for a typical living room. Note that some means of general illumination is essential. This general lighting may be provided by ceiling or wall fixtures, by lighting in coves, valances, or cornices, or by portable lamps. Coves, valances, etc. are illustrated subsequently. Lighting outlets should be wall-switch controlled, in locations appropriate to the lighting method selected. These provisions apply also to sun rooms, enclosed porches, television rooms, libraries, dens, and similar areas. Installation of outlets for decorative lighting accent, such as picture illumination and bookcase lighting, is also desirable.

Typical cove-lighting arrangements are depicted in Figure 6-2. This is a form of indirect lighting. A living room should be provided with convenience outlets so that no point along the floor line in any usable wall space is more than 6 feet from an outlet in that space. Where the installation of windows extending to the floor prevents meeting this requirement by the use

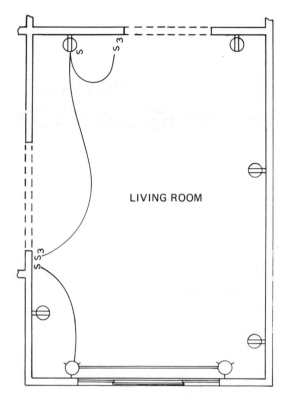

Figure 6-1 Lighting provisions for a typical living room

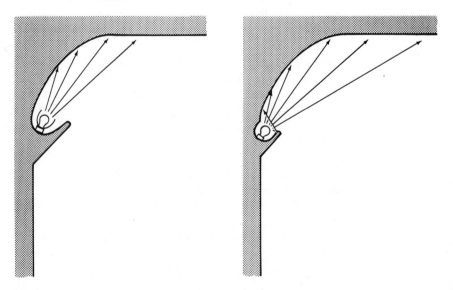

Figure 6-2 Examples of cove lighting arrangements

Figure 6-3   An example of valance lighting

of ordinary convenience outlets, equivalent facilities should be installed by utilizing other appropriate means. For example, a floor outlet could be installed.

If, instead of fixed lighting, general illumination is provided by portable lamps, two convenience outlets or one plug position in two or more split-

Figure 6-4   An example of cornice lighting

receptacle convenience outlets should be wall-switch controlled. In other words, a split receptacle has one outlet wired to a switch and the other outlet is connected directly across the line. In the case of switch-controlled convenience outlets, a split receptacle permits the operation of units that are normally not switched, such as electric clocks. This is also a desirable feature for operation of radio, television, and high-fidelity equipment.

When the building construction permits, it is desirable that one convenience outlet be installed flush in a mantel shelf. In addition, a single convenience outlet should be installed in combination with the wall switch at one or more of the switch locations for the use of a vacuum cleaner or other portable appliances in the living room. Outlets for supplying decorative lighting, clocks, radios, and so on may be provided in book cases and other suitable locations. It is also advisable to install one outlet for a room air conditioner wherever a central air-conditioning system is not being planned.

6–2 DINING AREA REQUIREMENTS

Figure 6-5 shows lighting provisions for a typical dining room. Each dining room, or dining area combined with another room, or breakfast nook, should have at least one lighting outlet with a wall switch. It is preferable to install such outlets over the location of the dining or breakfast table, to

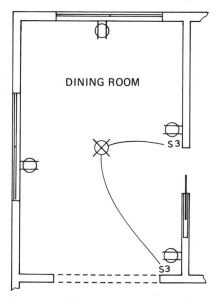

Figure 6-5 Lighting provisions for a typical dining room

provide direct illumination of the area. Convenience outlets should be placed so that no point along the floor line is more than 6 feet from an outlet in that space. When the dining or breakfast table is to be placed against a wall, one of these outlets should be placed at the table location, just above table height. Where open counter space is to be built in, an outlet should be provided above counter height for the use of portable appliances. Convenience outlets in dining areas should be of the split-receptacle type for connection to appliance circuits.

6–3 BEDROOM LIGHTING REQUIREMENTS

Figure 6-6 shows lighting provisions for typical bedrooms. Good general illumination is required in a bedroom. This should be provided by a ceiling fixture or from lighting in valances, coves, or cornices. Wall-switch controlled outlets should be provided in locations that are appropriate for the lighting method that is selected. Light fixtures over full-length mirrors, or a light source at the ceiling located in the bedroom and directly in front of clothes closets, may serve as general illumination. Master-switch control in the

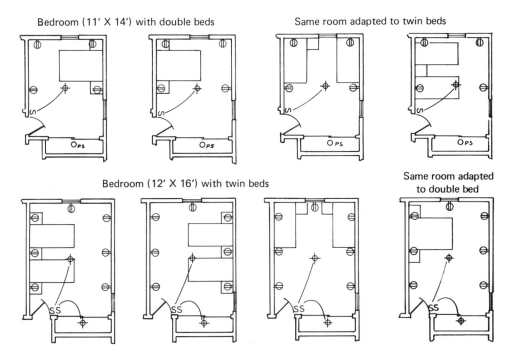

Figure 6-6 Lighting provisions for typical bedrooms

master bedroom, as well as at other desirable points in the home, may be installed for selected interior and exterior lights.

Outlets should be placed so that there is a convenience outlet on each side and within 6 feet of the center line of each bed location. Additional outlets should be placed so that no point along the floor line in any usable wall space is more than 6 feet from an outlet in that space. It is preferable that convenience outlets be placed only 3 to 4 feet from the center line of the bed locations. The popularity of bedside radios, bed lamps, and electric blankets makes more plug-in positions at bed locations essential. Triplex or quadruplex (three- or four-unit) convenience outlets are desirable at these locations. At some conveniently accessible place, a receptacle outlet should be provided for a vacuum cleaner, floor polisher, or other portable appliances.

Installation of one heavy-duty, special-purpose outlet in each bedroom for connection of a room air conditioner is recommended. Such outlets may also be used for portable space heaters during cool weather in climates where a small amount of local heat is sufficient. Figure 6-7 shows plans for both double- and twin-bed arrangements, and also their application where more than one probable bed location is available within the room. Wall brackets are often undesirable for bedroom illumination because they could tend to limit furniture arrangements. Ceiling fixtures may be of the semi-direct or totally indirect types. Three places that require special lighting are the vanity dresser, bed, and the boudoir chair or chaise lounge. Each closet that is more than 2 feet deep should have a light installed just inside above the door.   .

6–4 BATHROOM AND LAVATORY LIGHTING

Figure 6-7 shows lighting provisions for a typical bathroom. Illumination of both sides of the face when at the mirror is essential. Several methods are ordinarily employed to obtain good bathroom lighting. Lighting outlets should be installed to best suit the method selected, being careful to avoid a single concentrated light source, either on the ceiling or the side wall. All lighting outlets should be wall-switch controlled. A ceiling outlet located in line with the front edge of the basin will provide improved lighting at the mirror, general room lighting, and safety lighting for combination shower and tub.

When more than one mirror location is planned, equal consideration should be given to the lighting in each case. It is helpful to install a switch-controlled night light for convenience and safety. Where an enclosed shower stall is planned, an outlet for a vapor-proof lighting fixture should be installed, controlled by a wall switch outside the stall. One outlet is required near the mirror, from 3 to 5 feet above the floor. An outlet should be installed at

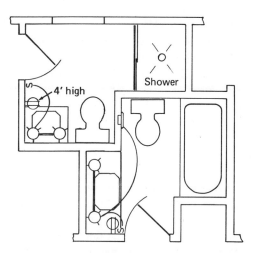

Figure 6-7 Lighting provisions for a typical bathroom

each separate mirror or vanity space and also at any space that might be used with an electric towel dryer, electric razor, or other utensil.

A receptacle which is part of a bathroom lighting fixture should be rated for 15 amperes and wired with at least 15-ampere wires. Special-purpose outlets should also be considered. For example, it is desirable that each bathroom be equipped with an outlet for a built-in type space heater and an outlet for a built-in ventilating fan with wall-switch control. Note that local building codes may prohibit the use of wall-switch control—in such

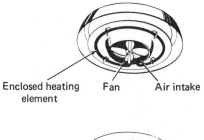

Enclosed heating element   Fan   Air intake

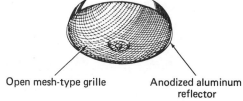

Open mesh-type grille   Anodized aluminum reflector

Figure 6-8 Ceiling-type radiant heater with a fan

cases, the ventilating fan must be controlled by the bathroom light switch. In some areas, installation of a bathroom ventilating fan is not required legally. Figure 6-8 depicts a ceiling-type radiant heater with a fan. Heat lamps may be installed instead of a heater. Note that the fan in Figure 6-9 does not serve the purpose of a ventilating fan. In other words, a heater fan provides downward air current, whereas a ventilating fan provides an upward air current that flows into a vent pipe.

6–5 KITCHEN LIGHTING PROVISIONS

Outlets should be provided both for general illumination and for lighting at the sink, as shown in Figure 6-9. These lighting outlets should be wall-switch controlled. Illumination is needed for the work areas, sink, range, counters, and tables. Undercabinet lighting fixtures within easy reach may have local-switch control. Consideration should also be given to outlets for inside lighting of cabinets. An ample number of convenience outlets is desirable. One outlet is provided for a refrigerator. In addition, one outlet should be installed for each 4 linear feet of work-surface frontage, with at least one outlet to serve each work surface. These work-surface outlets should be located approximately 44 inches above the floor level.

If a planning desk is to be utilized in the kitchen, one outlet should be installed to serve this area. An outlet should be provided for the table space, just above table level. In addition, an outlet is recommended at any wall space that may be used for ironing or for an electric roaster. Convenience outlets in the kitchen, other than that for the refrigerator, should be of the split-receptacle type that are connected into separate circuits. In other words,

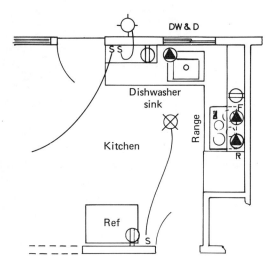

Figure 6-9 Typical kitchen lighting provisions

one of the outlets in a duplex receptacle is powered from one branch circuit, and the other outlet is powered from another branch circuit. Neither of the outlets is switch-controlled in this situation. Details are explained subsequently. Special-purpose outlets are also needed, with one outlet each for a range and a ventilating fan. One or more outlets for a dishwasher or a food waste disposer will be required, if the necessary plumbing facilities are installed. A clock outlet should be provided where it will be easily visible from all parts of the kitchen—a recessed receptacle with a clock hanger is preferred. If a food freezer is to be located in the kitchen, a suitable outlet should be installed.

6–6 CLOSET LIGHTING

One outlet should be provided for each closet. Where shelving or other conditions make the installation of lights within closets impractical, lighting outlets should be provided elsewhere so that adjoining spaces will illuminate the interiors of the closets. Wall switches should be installed near closet doors; otherwise, door-type switches that operate automatically may be provided. In any case, pull chains are undesirable, as they are comparatively difficult to find in a dark closet.

6–7 HALL LIGHTING

Wall-switch controlled lighting outlets should be installed for good illumination of the entire hall area. Particular attention should be given to irregularly shaped areas. Passage halls, reception halls, vestibules, entries, foyers, and similar areas also require careful planning. A night light with switch control should be installed in any hall that provides access to bedrooms.

Convenience outlets should be provided with one outlet for each 15 linear feet of hallway, measured along the center line. If a hall has over 25 square feet of floor area, at least one outlet should be installed. In reception halls and foyers, convenience outlets should be spaced so that no point along the floor line in any usable wall area is more than 10 feet from an outlet in that space. It is desirable that at one of the switch outlets a convenience receptacle be provided for connection of a vacuum cleaner, floor polisher, or other appliance.

6–8 STAIRWAY LIGHTING

Wall or ceiling outlets should be installed in stairways to provide adequate illumination of each stair flight. These outlets require multiple-switch control at the head and foot of the stairway, arranged so that full illumination may be turned on from either floor. It is desirable to install a switching system that will extinguish lights in halls furnishing access to bedrooms without interfering with ground-floor usage. These provisions are applicable to any stairway with finished rooms at both ends. Whenever possible, switches are grouped together and are not located so that a person might fall when reaching for a switch close to steps. Convenience outlets are desirable, and at in-between landings of a large area, an outlet is generally installed for a decorative lamp, night light, vacuum cleaner, or other appliance.

6–9 RECREATION ROOM

Figure 6-10 shows typical lighting provisions for a recreation room. Some means of general illumination is essential. This lighting may be provided by ceiling or wall fixtures, or by lighting in coves, valances, or cornices. Wall-

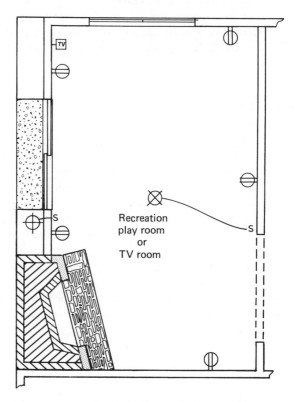

Figure 6-10 Typical lighting provisions for a recreation room

switch controlled lighting outlets are customarily provided in locations that are appropriate for the lighting method selected. This method depends on the type of major activities for which the room is planned. Convenience outlets are installed so that at any usable wall space no point along the floor line is more than 6 feet from an outlet in that area. It is desirable that one convenience outlet be installed flush in the mantel shelf, if construction permits. Outlets for a clock, radio, television, hi-fi, motion-picture projector, ventilating fan, and the like, are installed where they will be needed.

6–10 UTILITY ROOM AND LAUNDRY AREA

Figure 6-11 depicts typical lighting provisions for a utility room. Lighting outlets are placed to illuminate the furnace area, and also a work bench, if planned. One outlet is generally wall-switch controlled. A convenience outlet, preferably near the furnace location or near a work-bench location, will be needed. In addition, a special-purpose outlet for electrical equipment associated with furnace operation is required. It is evidence of poor planning to skimp on utility-room wiring. Correction of inadequate wiring at some future time is comparatively difficult and costly. Special-purpose outlets and circuits are explained subsequently.

Figure 6-12 shows typical laundry lighting provisions. In the case of complete laundries, outlets are planned to provide proper illumination of work areas such as laundry tubs, sorting table, washing, ironing, and drying centers. At least one lighting outlet in a laundry room needs to be wall-switch

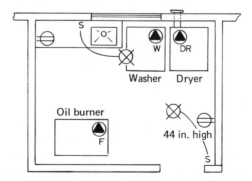

Figure 6-11 Typical laundry lighting provisions

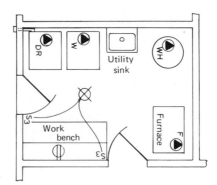

Figure 6-12 Typical lighting provisions for a
utility room

controlled. For laundry trays in an unfinished basement, one ceiling outlet
centered over the trays is adequate. In general, pull chains or key switches
are less desirable than wall-switch control of all laundry lighting. At least
one convenience outlet will be needed. Depending upon the wiring plan, one
of the special-purpose outlets may meet this requirement, if suitably located.
This convenience outlet will be utilized for appliances such as a laundry
hot plate, sewing machine, and so on. A laundry-area convenience outlet will
accommodate heavier appliances if it is of the split-receptacle type (each
outlet of the duplex receptacle supplied by a separate branch circuit).

Special-purpose outlets are planned with one outlet for each of the following appliances: automatic washer, hand iron or ironer, and clothes dryer.
Installation of outlets for a ventilating fan and a clock is desirable. If an electric water heater is planned, the requirements should be determined by
consulting with the local public-utility company. Note that in some areas,
a separate meter might be provided with a time switch, so that the heater
can be operated at a reduced rate. In other words, the switch automatically
opens the circuit during peak-load times, and there will be several hours
during the day when the water heater is turned off.

6–11 BASEMENT LIGHTING

Typical lighting provisions for a basement are shown in Figure 6-13.
Lighting outlets should be located to illuminate designated work areas or
equipment locations, such as furnace, pump, work bench, and the like.
Additional outlets are often needed near the foot of the stairway, in each
enclosed space, and in open spaces so that each 150 square feet of open area
is adequately served by a light in that area. In unfinished basements, the

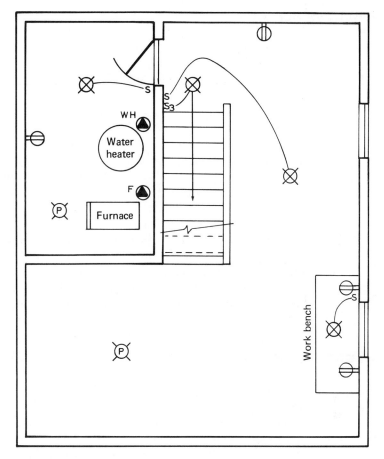

Figure 6-13 Typical lighting provisions for a basement

light at the foot of the stairs is customarily wall-switch controlled near the head of the stairs. However, other lights are often pull-chain controlled.

In basements with finished rooms, plus garage space, or with other direct access to outdoors, the foregoing stairway lighting provisions are applicable. When a basement will be infrequently visited, a pilot light should be installed in conjunction with a switch at the head of the stairs, to ensure against waste of electricity. At least two convenience outlets should be provided. If a work bench is planned, an outlet will be required at this location. Basement convenience outlets are useful near a furnace, play area, basement laundries, dark rooms, hobby areas, and for appliances such as a dehumidifier, portable space heater, and so on. A special-purpose outlet is needed for electrical equipment used in connection with furnace operation. An outlet for a food freezer is also desirable.

6–12 ACCESSIBLE ATTIC

An accessible attic requires one outlet for general illumination, wall-switch controlled from the foot of the stairs. In case no permanent stairs are installed, this lighting outlet may be pull-chain controlled, provided that it is located over the access door. If an unfinished attic is planned for later division into rooms, the attic-lighting outlet needs to be switch-controlled at the top and at the bottom of the stairs. An outlet should be provided for each enclosed space. These provisions are applicable to unfinished attics. On the other hand, ordinary room requirements apply to attics with finished rooms. Note that it is desirable to install a pilot light in combination with a switch that controls the attic light.

One convenience outlet is needed for general use in an attic. If an open stairway leads to future attic rooms, it is good planning to provide a junction box with direct connection to the distribution* panel for future extension to convenience outlets and lights when the rooms are finished. A convenience outlet in the attic is desirable for providing additional lighting in dark corners and also for use of a vacuum cleaner and its accessories. Installation of a special-purpose outlet that is multiple-switch controlled from desirable points throughout the house, is also desirable for operation of a summer cooling fan.

6–13 PORCH LIGHTING

Figure 6-14 shows typical porch-lighting provisions. Each porch, breezeway, or other similarly roofed area of more than 75 square feet in floor area requires a lighting outlet that is wall-switch controlled. Large or irregularly shaped areas may require two or more lighting outlets. Multiple-switch control is generally installed at entrances when the porch is used as a passage between the house and garage. One convenience outlet, weatherproof if exposed to moisture, is needed for each 15 feet of a wall along a porch or breezeway. It is desirable that all such outlets be controlled by a wall switch inside the door. Note that the split-receptacle convenience outlet shown in Figure 6-14 is intended to be connected to a three-wire branch circuit. This area is considered as an outdoor dining area. Three-wire branch circuits are equivalent to a pair of basic branch circuits, and are explained in greater detail subsequently.

6–14 TERRACES AND PATIOS

Installation of an outlet on the building wall or on a post centrally located in a terrace or patio area is desirable for a source of general lighting. Such outlets are customarily wall-switch controlled just inside the house door

---

*A distribution panel may be a service panel, or it may be an auxiliary panel with feeder wires running to the service panel.

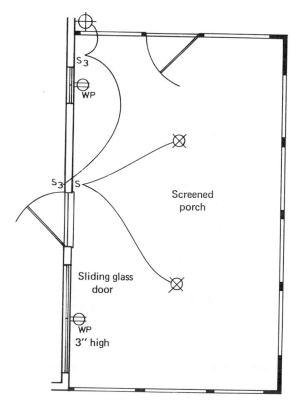

Figure 6-14 Typical porch-lighting provisions

opening onto the area. One weatherproof convenience outlet will be needed for each 15 linear feet of house wall along a terrace or patio. These outlets are generally located at least 18 inches above grade line and are wall-switch controlled from inside the house. Garden lights may be utilized for decoration, utility, or both. There is a spike type of light that is a portable unit. Garden lights can also be installed permanently on 2-foot or 4-foot posts.

6–15 EXTERIOR ENTRANCES

An exterior entrance should be provided with one or more lighting outlets, as the building plan calls for, wall-switch controlled, at front and trade entrances. Where a single wall outlet is desired, location on the latch side of the door is preferable. It is desirable that lighting outlets, wall-switch controlled, be installed at other entrances also. Note that the chief lighting requirements are illumination of steps leading to the entrance and of faces of people at the door. Outlets in addition to those at the door are often desirable for post lights to illuminate terraced or broken flights of steps on long approach walks. These outlets need to be wall-switch controlled inside the house entrance. A weatherproof outlet, preferably near the front entrance,

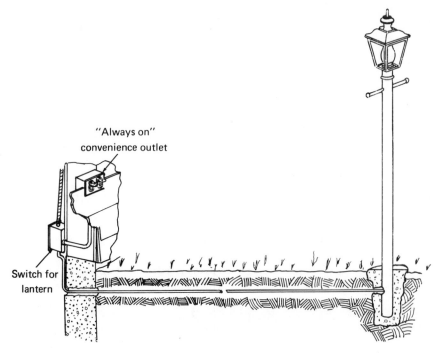

Figure 6-15   Installation of a post lantern and convenience outlet

should be located at least 18 inches above grade. It is desirable that this outlet be controlled by a wall switch inside the entrance for convenient operation of outdoor decorative lighting. Wall-switch controlled outlets are also needed to serve garden appliances.

Light clusters are often used to illuminate exterior entrances. Post or porch lanterns, shown in Figure 6-15, ensure well-lighted driveways, sidewalks, and doorways. Note that post lanterns are wired underground with dual-purpose plastic cable. In some locations, local codes may require underground conduit, buried at least 2 feet deep. Post lanterns are usually switch-controlled from inside the house. If desired, the manual switch may be supplemented by a photoelectric switch that automatically turns the lights on at sunset and turns them off at sunrise. Post lanterns may also be time-switch controlled, if desired.

6–16 FARM BUILDINGS

Figure 6-16 depicts a typical dairy barn wiring layout. One lighting outlet is required for every three stalls or less, on the center line of the alley, with wall-switch control for the litter alley. An outlet should be installed every 15 feet or less, on the center line of the feed alley, with separate wall-switch control. Special-purpose outlets may be needed for ventilating fans. In a loose-housing barn, box stalls and pens are provided with a light outlet for each

150 feet of open pen area, and one ceiling outlet for each bull, maternity, or calf pen, with wall-switch control. Convenience outlets are needed every 20 feet for portable milkers, clippers, and other appliances.

One light outlet is needed in front of every three cows in the milking room, and one outlet behind every two cows, with wall-switch control. A convenience outlet should be provided for every five cows or less. In the milkhouse, one light for each 100 square feet will provide good general illumination. Each work area requires a convenience outlet for motor-driven devices of less than 1/3 horsepower, and for plug-in type water heaters that consume 250 to 1,000 watts. Special-purpose outlets are often needed for a can-type milk cooler, bulk-type cooler, utensil sterilizer, fan, heating equipment, or water heater. These may require either 120- or 240-volt circuits. Outlets are mounted at a sufficiently high level so that they will not be splashed.

Beef cattle barns require a wall-switch controlled light outlet for every 150 square feet of open pen area. If a feed alley is utilized in either a beef cattle or sheep barn, a light outlet is customarily installed each 15 feet. In hog and farrowing houses, one wall-switch controlled light outlet should be installed for each pair of hog pens. These outlets are located over the partition lines. Low houses may require a double row of ceiling lights, with one outlet for each pair of pens. In cattle barns, an outlet is needed at each location where equipment is used. Farrowing pens in sheep barns or hog houses require convenience outlets for brooders and water warmers.

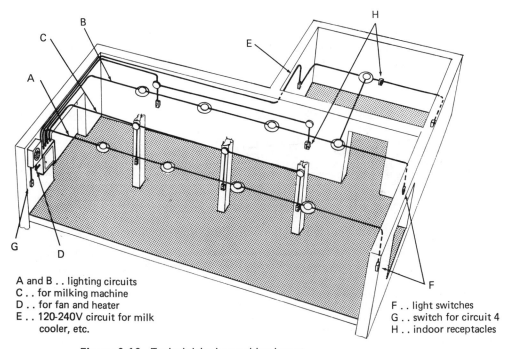

A and B . . lighting circuits
C . . for milking machine
D . . for fan and heater
E . . 120-240V circuit for milk
    cooler, etc.

F . . light switches
G . . switch for circuit 4
H . . indoor receptacles

Figure 6-16 Typical dairy barn wiring layout

It is good practice to install three-wire 120- to 240-volt circuits for heat lamps, connecting alternate outlets to opposite sides of the circuit in order to obtain good load balance. This topic is discussed is greater detail later. All barns need to have a special-purpose outlet for connection of a heating cable or other arrangement to keep the water supply from freezing. In turn, several branch circuits will be required. As noted before, each circuit may be rated for 15 or 20 amperes. An outlet for a heat lamp must supply 1.5 amperes or 175 watts, in a typical installation. Therefore, if 20-ampere circuits are utilized, each circuit can operate approximately a dozen heat lamps. Branch circuiting is discussed more extensively in Chapter 9.

Figure 6-17 depicts a typical poultry house wiring layout. Both dim and bright light circuits are required to obtain maximum egg production. For morning lighting, it is desirable to have one outlet for each 200 square feet of floor area. If the house or pen is 20 feet deep, at least one row of outlets should be placed 10 feet apart midway between the front of the house and the dropping board. A time-switch control is required for the outlets. For morning and evening lighting, bright lights are installed as noted above. Dim lights are installed on a separate circuit. One outlet for each 400 square feet of floor space is standard installation practice. These outlets should be lined up slightly in back of the bright lights. Ten-watt bulbs are utilized, with time-switch control.

All-night lighting is provided by means of a 10- or 15-watt light for every 20 square feet. Two lights may be employed to make certain of ample illumination when one lamp burns out. They are preferably installed over feed and water areas, with switch control for each pen. In addition, outlets are necessary for motor-driven feeders and for warming poultry drinking water in cold locations. One outlet is required for every 400 feet of floor area, with

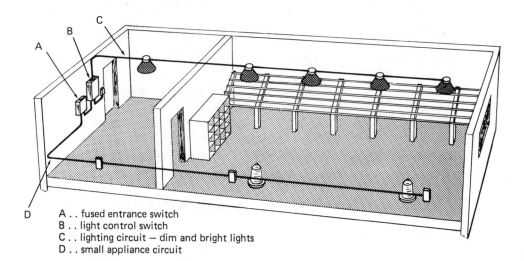

A . . fused entrance switch
B . . light control switch
C . . lighting circuit – dim and bright lights
D . . small appliance circuit

Figure 6-17 Typical poultry house wiring layout

one outlet for any pen that is at least 3 feet high. Special outlets may be needed for fans, conveyors, hoists, or elevators. In brooder houses, one outlet is needed for each pen or house. Also, special-purpose outlets should be installed on the ceiling for each 1,000 watts of hover-type or four-lamp infrared brooder. When a number of units are utilized, three-wire split-circuit 120- to 240-volt wiring should be employed. For battery brooders, incubators, egg storage and handling rooms, or poultry cleaning and dressing rooms, one outlet is needed for every 200 square feet of floor space. Convenience outlets, special outlets, and incidental circuits may also be required.

A farm shop requires one outlet for each 200 square feet for general illumination. One outlet should also be provided for each fixed unit of equipment, such as a grinder or a saw. It is advisable to plan for at least one outlet for every 10 feet of bench space, with wall-switch control. A convenience outlet should be provided for each 5 feet of bench space, and one outlet for every fixed unit of less than 1/3 horsepower, such as a forge or drill. A motor rated at 1/2 horsepower or more should be provided with a special-purpose 240-volt outlet. Farm welders typically draw 37-1/2 amperes at 240 volts, and utilize a 50-ampere three-wire outlet. Grounding circuits and shock protection are vital in rural wiring systems, owing to the high moisture level that often prevails.

6–17 ELECTRICAL ESTIMATING

An electrical wiring project involves both material and labor costs. Costs of wire, boxes, switches, outlet receptacles, fixtures, GFCI's, and service entrance equipment can be calculated after the project has been planned. Labor costs depend on the prevailing wage scale, efficiency of the electrician, and whether part or all of the labor might be done by the property owner. For example, a highly skilled electrician might rough-in a typical tract house in one day. Then, after the walls and ceilings have been finished, the electrician might complete the installation of switches, receptacles, wall plates, and fixtures in one more day. On the other hand, a moderately skilled electrician might take twice as long to do the same work. An electrical contractor knows how much time will be required for an electrician in his employ to complete a typical project.

Instead of making a detailed breakdown on residential wiring projects, the usual procedure is to estimate in terms of the number of boxes that are to be installed, plus certain supplementary costs. As an illustration, if the wage scale for electricians is $11 per hour, an electrical contractor will charge about $12 for installation of each box. A supplementary charge of $125 is typically made for service installation. However, if an underground service installation is to be made, this supplementary charge is increased accordingly. Supplementary charges are also made for each 240-volt outlet, because material costs are somewhat greater than for 120-volt outlets. In the case of a dryer or an electric range, a separate charge is made for cable, if the run is over 10 feet. In other words, this type of cable is comparatively heavy and costly.

In estimating a wiring project, a contractor charges an amount that allows about $18 per hour for labor, assuming that the electricians' wage scale is $11 per hour. In turn, the contractor realizes about $7 per hour profit on the electrician's labor. Note that a contractor also takes a profit on material, inasmuch as he purchases fixtures, for example, at wholesale cost and resells them at the retail price. Many electrical contractors operate retail lighting and appliance stores in combination with their contracting business. Note in passing that electricians' wage scales in metropolitan areas are usually higher than in rural areas.

6–18 HOW TO MAKE ELECTRICAL LAYOUTS

Before the circuiting for a wiring system can be drawn, we must know where the lighting fixture outlets, light switches, and convenience outlets will be located in a room. Figure 6-6 shows an outlet and switch layout for a small bedroom. We proceed as follows:

1. Draw the room plan to scale. It is helpful to use a sheet of graph paper and let one square equal 1 foot.

2. In this example, one ceiling outlet is provided for a lighting fixture. This outlet is represented by a circle.

3. A switch outlet is conveniently located on the wall near the entrance to the room. This switch is represented by S, and a dotted line is drawn from the switch to the ceiling outlet. This dotted line indicates that the switch will control the ceiling fixture.

4. Appliance (or portable lamp) outlets are located on three walls of the room. These receptacle outlets are of the duplex type that permits two different devices to be plugged in at the same time. A duplex outlet is represented by a circle with two lines running through the circle and meeting the wall at the point where the outlet is to be installed.

6–19 NEW WORK AND OLD WORK

An electrical installation will be classified as either new work or old work. New work consists of wiring in buildings that are under construction. On the other hand, old work involves alterations or additions to wiring systems in completed buildings. In some cases, a completed building in a remote area might not have any wiring—if a wiring system is to be installed, the project is classified as old work. Note that old work is also called old-house wiring. Local electrical codes may require all new work to employ conduit. Note that alterations in old work are made easier if conduit was previously installed because wires can be pulled out easily and other wires can be "fished" through the conduit. In general, new work can be done quicker and with less difficulty than old work.

STUDENT EXERCISE

This exercise provides familiarization with the standard electrical symbols used in architectural drawings. With reference to Table 6-1, and Figures 6-1, 6-5, 6-6, 6-8, 6-10, 6-11, 6-12, 6-13, 6-14, 6-15, and 6-17, identify each of the electrical symbols in the figures.

Table 6-1 Graphical electrical symbols for residential wiring plans

General Outlets

Lighting Outlet

Ceiling Lighting Outlet for recessed fixture (Outline shows shape of fixture.)

Continuous Wireway for Fluorescent Lighting on ceiling, in coves, cornices, etc. (Extend rectangle to show length of installation.)

Lighting Outlet with Lamp Holder

Lighting Outlet with Lamp Holder and Pull Switch

Fan Outlet

Junction Box

Drop-Cord Equipped Outlet

Clock Outlet

To indicate wall installation of above outlets, place circle near wall and connect with line as shown for clock outlet.

Convenience Outlets

Duplex Convenience Outlet

Triplex Convenience Outlet (Substitute other numbers for other variations in number of plug positions.)

Duplex Convenience Outlet — Split Wired

Duplex Convenience Outlet for Grounding-Type Plugs

Weatherproof Convenience Outlet

Multi-Outlet Assembly (Extend arrows to limits of installation. Use appropriate symbol to indicate type of outlet. Also indicate spacing of outlets as X inches.)

Combination Switch and Convenience Outlet

Combination Radio and Convenience Outlet

Floor Outlet

Range Outlet

Special-Purpose Outlet. Use subscript letters to indicate function. DW-Dishwasher, CD-Clothes Dryer, etc.

Switch Outlets

S Single-Pole Switch

S_3 Three-Way Switch

S_4 Four-Way Switch

S_D Automatic Door Switch

S_P Switch and Pilot Light

S_{WP} Weatherproof Switch

S_2 Double-Pole Switch

Low-Voltage and Remote-Control Switching Systems

$\underline{S}$ Switch for Low-Voltage Relay Systems

$\underline{MS}$ Master Switch for Low-Voltage Relay Systems

O_R Relay—Equipped Lighting Outlet

— — — —   Low-Voltage Relay System Wiring

Auxiliary Systems

Push Button

Buzzer

Bell

Combination Bell-Buzzer

Chime

Annunciator

Electric Door Opener

Maid's Signal Plug

Interconnection Box

Bell-Ringing Transformer

Outside Telephone

Interconnecting Telephone

Radio Outlet

Television Outlet

Miscellaneous

Service Panel

Distribution Panel

Switch Leg Indication. Connects outlets with control points.

Special Outlets. Any standard symbol given above may be used with the addition of subscript letters to designate some special variation of standard equipment for a particular architectural plan. When so used, the variation should be explained in the Key of Symbols and, if necessary, in the specifications.

STUDENT EXERCISE

With reference to the electrical wiring layout in Figure 3-8, calculate the approximate cost of the project, using the cost figures given in Section 6-17. For the purpose of rough estimation, compute the cost of each 240-volt outlet at 50% more than for a 120-volt outlet.

QUESTIONS AND EXERCISES

1.   What are some of the important factors involved in planning a wiring project?
2.   How does a split receptacle differ from a basic duplex receptacle?
3.   How are split receptacles subdivided into types?
4.   Make a drawing of a typical bedroom and indicate convenience outlets, lights, and switches.
5.   Draw a complete wiring diagram for a kitchen.
6.   What is some of the electrical equipment utilized in a bathroom that requires special wiring?
7.   How should closets be illuminated?
8.   What number of convenience outlets would you install in a hallway?
9.   What kind of electrical equipment and appliances would be utilized in a basement?
10.   How are exterior convenience outlets planned, and what are some of the special requirements?
11.   Draw a wiring diagram for a typical dairy barn.
12.   What are the electrical installation requirements for a farm shop?

7 WIRING PRACTICES

7-1 GOOD PRACTICES AND CODES

Many of the considerations that enter into work plans are absolutely necessary, being required by the National Electrical Code and prevailing local codes. These requirements represent the simplest work plans possible, and they are usually added to by considerations of good practices. As an illustration, although the NEC specifies the smallest number of convenience outlets that must be installed in an area, good practices will often call for additional outlets at various locations. Where required on a basis of linear or square-foot measure, the number of outlets should be determined by dividing the total linear or square footage by the required distance or area, and the number so determined should be increased by one if a major fraction remains. For example:

> One outlet is required for each 150 square feet.
> Total square feet in a given area equals 390.
> 390 divided by 150 equals 2.6
> Utilize three outlets.

In any situation where an outlet is located so as to satisfy two different provisions from the viewpoint of adequacy, only one outlet need be installed at that location. In such instances, particular attention should be paid to any required wall-switch control, because additional switching may be necessary. For example, a lighting outlet in an upstairs hall may be located at the head of the stairway, thus satisfying both a hall lighting outlet and a stairway lighting outlet requirement with a single outlet. Stairway requirements will necessitate multiple-switch control of this lighting outlet at the

Figure 7-1 Examples of Underwriters' Laboratories seals (Courtesy Underwriters' Laboratories Inc.)

head and foot of the stairway. In addition, this multiple-switch control rule, when applied to the upstairs hall, may require a third point of control elsewhere in the hall because of its length. In summary, a single lighting outlet satisfied two requirements in the foregoing example, provided that the switching circuit was suitably elaborated.

If defective or shoddy materials are utilized in an electrical wiring system, they can present personal and property hazards. Therefore, the National Board of Fire Underwriters has established the Underwriters' Laboratories, Inc. Any electrical manufacturer who wishes to obtain U-L approval for his products may submit samples to a U-L testing laboratory. If a product is approved, the manufacturer is permitted to place the U-L seal on his product. Typical U-L seals are illustrated in Figure 7-1. When a product is listed by the U-L, the user knows that it meets the minimum required safety standards, when installed or applied in the manner specified by the National Electrical Code or, if not covered by the NEC, in accordance with good practices.

Although most good-quality electrical products are U-L listed, there are a few exceptions. An occasional manufacturer who has been in business for many years, and whose trademark is recognized as a symbol of quality, may not seek U-L listing. Also, only products that could impose a safety hazard are processed by the Underwriters' Laboratories. Thus, the U-L seal provides useful guidance for inexperienced purchasers of electrical products.

It is impractical to study the National Electrical Code as if it were an ordinary textbook, because the various sections present facts and regulations without progressive development. Therefore, you should study the National Electrical Code by planning installations and working out problems. At various points in the planning procedure, questions will arise concerning NEC requirements. Then, if you refer to the regulations in the code, they will become meaningful to you, and you will remember them. Similarly, if you are working on a wiring problem, questions concerning spacing of outlets, conductor sizes and insulation, and percentage ratings of maximum load demands will occur. You should refer to the NEC when such questions arise. As you proceed with your studies in this manner, you will progressively learn what you need to know of the code, and you will remember what you have learned.

7–2 WIRING THE SERVICE ENTRANCE

Figure 7-2 shows an arrangement for a service entrance. In most locations, the wiring up to the service head above the watt-hour meter is done by the utility company. Note that the service drop wires are overhead wires. Service-entrance wires run from the meter to the service switch. In this example, the installation is made with galvanized rigid conduit. Observe

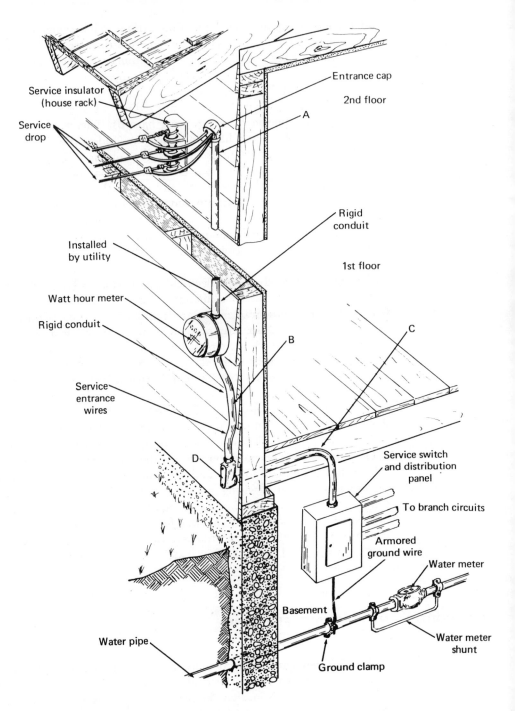

Service insulator
(house rack)

Service
drop

Entrance cap

2nd floor

A

Rigid
conduit

1st floor

Installed
by utility

Watt hour meter

Rigid conduit

Service
entrance
wires

B

C

D

Service switch
and distribution
panel

To branch circuits

Armored
ground wire

Water meter

Basement

Water meter
shunt

Water pipe

Ground clamp

Figure 7-2 Arrangement of a service entrance

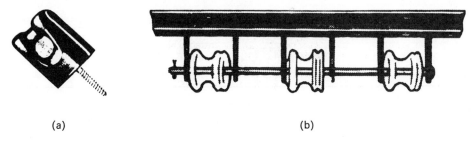

(a)   (b)

Figure 7-3 (a) Screw-type glazed porcelain insulator (b) service-wire holder

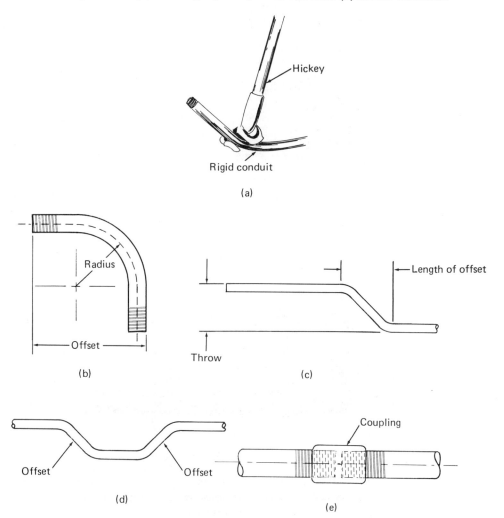

Figure 7-4 (a) Conduit is bent with a hickey (b) an ell bend (c) an offset bend (d) a saddle bend (e) threaded ends and coupling

149

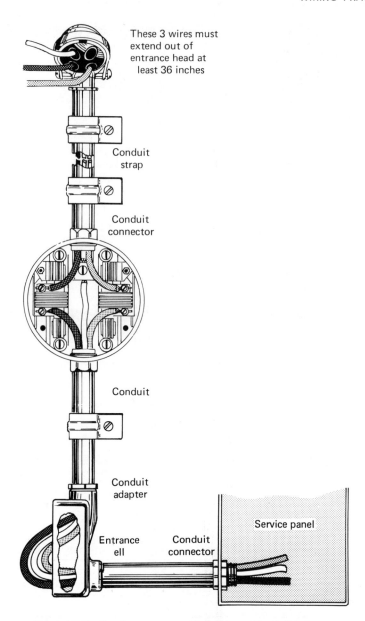

These 3 wires must
extend out of
entrance head at
least 36 inches

Conduit
strap

Conduit
connector

Conduit

Conduit
adapter

Entrance
ell

Conduit
connector

Service panel

Figure 7-5 Service entrance using thin-wall conduit

that the service insulators may be individual screw-type insulators or a
ganged service-wire holder or house rack, as depicted in Figure 7-3. A service-wire holder is also called a secondary rack.

Note that the conduit installation shown in Figure 7-2 starts with an
entrance cap, also called a service head or a weather head. It is mounted
higher than the service insulators, to prevent water from entering the conduit.

Table 7-1 Conduit sizes required in various situations

Conduit size	Conduit capacity
¾ inch	3 number 8 wires
1¼ inch	3 number 2 wires, 3 number 3 wires, 3 number 4 wires, 3 number 6 wires
1½ inch	3 number 1 wires
2 inch	3 number 1/0 wirs 3 number 1/0 wires, 3 number 3/0 wires

Entrance conduit runs cannot exceed 50 feet

In case the service drop wires need to be mounted higher than shown in Figure 7-2 to conform with applicable codes, the service insulators and the entrance cap would have to be installed on the roof of the building. As an illustration, in residential areas, the NEC requires that the service drop wires must be installed at least 12 feet above driveways or at least 10 feet above sidewalks.

As seen in Figure 7-2, the conduit is bent at points A, B, and C. It is bent with a hickey, or conduit bender, as depicted in Figure 7-4. Basic conduit bends are termed the ell, the offset, and the saddle bends. A saddle bend is shown at B in Figure 7-2, and an ell bend is depicted in C. Pipe clamps are used to secure the conduit in place. Note that conduit must be threaded with a stock and die, as in plumbing work. A threaded coupling is seen in Figure 7-4(e). Various types of conduit fittings are used in different situations. As an illustration, an access fitting or entrance ell is shown at D in Figure 7-2. It has a removable cover and is provided in order to pull the wires through the conduit and around the right-angle course after the conduit has been installed.

Thin-wall conduit is also used for service-entrance installations, as shown in Figure 7-5. A three-wire service employs a black, a white, and a red wire. As explained previously, the wire size that is utilized depends on the required ampacity. In turn, the size of conduit that is necessary depends on the gage of the wires, as tabulated in Table 7-1. Note that the service-drop wires must be at least three feet away from windows, doors, fire exits, or any opening in the building. Thin-wall conduit is secured by a metal strap or pipe clamp every 4 feet. Since thin-wall conduit is not threaded, connectors are utilized to secure the conduit at the meter and at the service panel. A conduit adapter is used to secure the conduit to the entrance ell. Note in passing that the

Figure 7-6 A metal fish tape (snake) is used for pulling wires

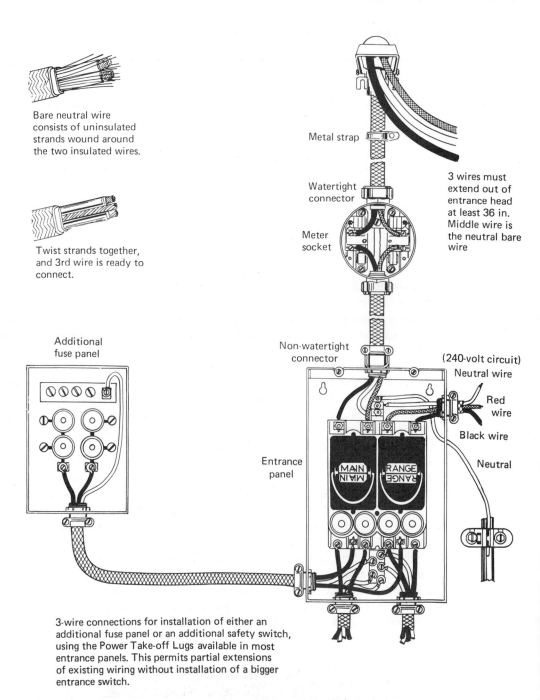

Bare neutral wire consists of uninsulated strands wound around the two insulated wires.

Twist strands together, and 3rd wire is ready to connect.

Metal strap

Watertight connector

Meter socket

3 wires must extend out of entrance head at least 36 in. Middle wire is the neutral bare wire

Additional fuse panel

Non-watertight connector

(240-volt circuit) Neutral wire

Red wire

Black wire

Neutral

Entrance panel

MAIN

RANGE

3-wire connections for installation of either an additional fuse panel or an additional safety switch, using the Power Take-off Lugs available in most entrance panels. This permits partial extensions of existing wiring without installation of a bigger entrance switch.

Figure 7-7 A service entrance using entrance cable

Figure 7-8 Typical sill plate. Opening is
sealed with a soft rubber compound

wires are inserted at the meter socket and then pushed or pulled through
the conduit. Electricians use a metal fish tape or snake, as depicted in Figure
7-6, for pulling wires.

Entrance cable is also used for installation of service entrances, as depicted
in Figure 7-7. Entrance cable is comparatively easy to work with; the installation procedure is basically the same as for thin-wall conduit. From the
meter, the cable is routed down the wall and through a hole drilled near
where the service panel is located inside. Note that the service panel should
be installed within 1 foot of the hole drilled through the wall. A sill plate
(Figure 7-8) is utilized to protect the cable where it enters the building. In
the example of Figure 7-7, the black wire from the cable is connected to
the left terminal of the main disconnect switch, and the red wire is connected
to the right terminal. Note that the bare neutral wire is connected to the
neutral strip, which in turn is grounded with a bare wire to a cold-water
pipe.

Observe in Figure 7-7 that there are four fuses in the entrance panel for
four branch circuits. Below each fuse, a black wire is connected to a terminal;
the white wires are connected to a neutral strip. A black wire and a white
wire provide a 120-volt branch circuit. Each black wire is fused with a 15-
amp or a 20-amp fuse, depending upon the ampacity of the branch-circuit
wires. Note also the additional fuse panel depicted in Figure 7-7. It is connected to the two power-takeoff lugs between the two left-hand and between
the two right-hand fuses in the entrance panel. Observe that the white wire
is connected to the neutral strip. Either 120-volt or 240-volt branch circuits
can be supplied from the additional fuse panel. As explained in greater
detail subsequently, additional (add-on) fuse panels are generally installed
in an existing wiring system when the necessity arises for additional branch
circuits. An add-on fuse panel is also called a distribution panel.

When the roof of a building is so low that the service drop does not have
minimum clearance above ground, a service mast is installed, as shown in
Figure 7-9. Large-diameter service conduit is utilized for the mast, and
flashing is generally used to provide a watertight seal. Next, observe the

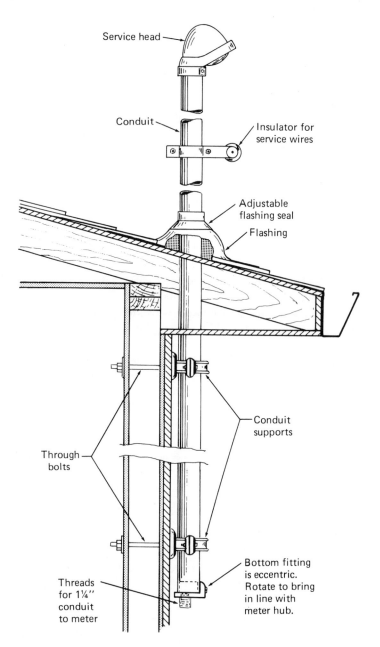

Figure 7-9 Typical mast installation

underground type of service entrance depicted in Figure 7-10. Codes usually permit the electrician to install the conduit and wires in the sidewalk handhole, but the wires must be connected to the service cable by the utility

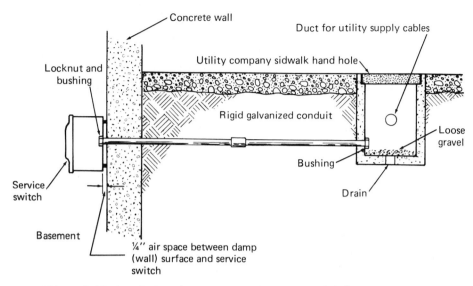

Figure 7-10 Installation of an underground service conduit (service lateral) to a sidewalk handhole

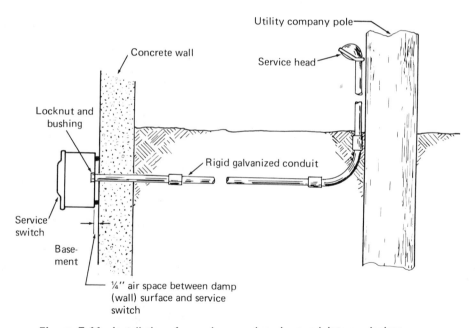

Figure 7-11 Installation of an underground service conduit to a pole riser

company. Again, when an underground service conduit runs to a pole riser, as depicted in Figure 7-11, the electrician installs the conduit up the pole for 8 feet and the installation is completed by the utility company.

7–3 WIRING INSIDE WALLS AND CEILINGS

Wiring inside of walls, as shown in Figure 7-12, is called concealed wiring. Conduit, armored cable, dual-purpose plastic cable, or nonmetallic sheathed cable (NMC) may be utilized, depending on applicable codes. Note that the studs are bored in the example of Figure 7-12 to pass the cable. Or, the studs may be notched, as described in greater detail later. When non-metallic sheathed cable is run up or down a stud or along a joist, the cable

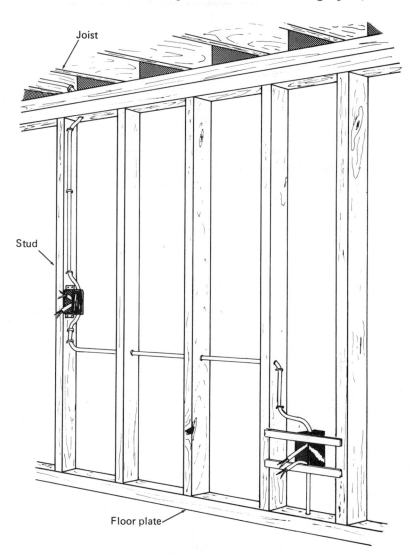

Figure 7-12 An example of concealed wiring

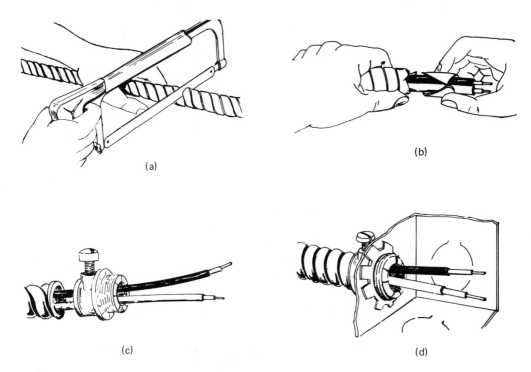

(a)

(b)

(c)

(d)

Figure 7-13 (a) Cutting with hacksaw (b) insertion of fiber bushing (c) application of connector (d) securing cable in box

should be strapped every 4-1/2 feet. It should also be strapped within 12 inches of every outlet or switch. Note, however, that in remodeling (old work), where the cable must be worked ("fished") behind the walls, strapping of cables is impractical and is not required. Observe that when a cable is stripped of its jacket at the end, at least 8 inches of insulated wire should be allowed for making connections.

Armored cable should also be strapped every 4-1/2 feet, except in old work, as noted above. A hacksaw is used to cut armored cable, as pictured in Figure 7-13. Note that the saw is held at a suitable angle to the cable for making a cut through one section of the spiral armor. Care must be taken not to cut the wires. Next, the armor is twisted on either side of the cut, to break the metal. At least eight inches of wire should be allowed for making connections. Since a sharp jagged edge is left by the saw at the end of the armor, the NEC requires insertion of a fiber bushing. As seen in Figure 7-13, when the armor is removed, the water-repellant paper is exposed. Note that the fiber bushing is inserted between the paper and the wires. In turn, the paper is removed and the connector (with locknut removed) is slipped over the wires and armor. Then, the fiber bushing is seated into the connector as far as it will go and the screw is tightened on the connector. If the armored

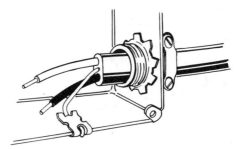

Figure 7-14 Installation of non-metallic cable into outlet box

cable has a bond wire, the wire is bent back against the armor and fastened to the screw of the connector. Finally, the connector is inserted into the knockout of the box, and the locknut is screwed up tightly in place.

Knockouts are usually removed with a screwdriver—a slot may be provided for prying out the metal disc. Or, the screwdriver may be given a blow with a pair of electrician's pliers to start the disc so that it can be grasped with the pliers and twisted out. If one of the knockouts is accidentally removed in

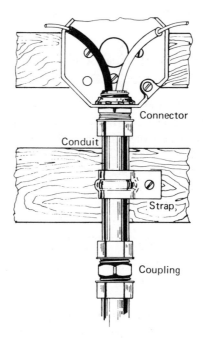

Figure 7-15 Thin-wall conduit with coupling, strap, and connector

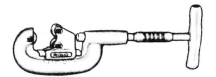

Figure 7-16 A pipe cutter used to cut EMT

a box, it can be plugged up with a closure washer. Nonmetallic cable is secured at the end with a double-screw squeeze-tight connector. This connector is inserted through a knockout, as depicted in Figure 7-14, and secured with a locknut. Note that local codes may require the use of a grounding wire that is connected to the box by means of a spring clip, as seen in Figure 7-14.

Installation of thin-wall conduit is not unduly difficult, although it is not as simple to install as flexible cable. On the other hand, it is easier to install thin-wall conduit than rigid conduit because EMT is much lighter and easier to cut and bend. Also, EMT is not threaded into connectors, as noted previously. EMT should be installed and connectors fitted to boxes before wires are inserted. Otherwise, serious difficulties would be encountered. Only steel boxes may be utilized with EMT. Note that this conduit is supplied in 10-foot lengths that are joined by couplings, as pictured in Figure 7-15. EMT is cut with a hacksaw and the ends are smoothed with a reamer and a file. A hickey is used to bend EMT, in the same manner as rigid conduit. EMT is cut with a pipe cutter such as illustrated in Figure 7-16.

If studs are not bored or notched to pass conduit, it should be supported every 10 feet with pipe straps when installing concealed wiring. Note that on exposed runs a pipe strap should be provided every 6 feet. After the wires are pulled through the conduit, 8 inches of wire are provided past the end of the conduit for making connections. Keep in mind that the NEC permits no more than four 90° bends in a run of conduit from one box to another. Boxes must be mounted at least 2 inches from any door frame. To assure uniform height of convenience outlets, electricians often use a 12-inch hammer handle as a guide in mounting outlet boxes above the floor level.

Figure 7-17 shows how thin-wall conduit is run to a ceiling outlet in new work, when a floor is to be laid overhead. Observe that the furring is cut to pass the EMT, and a 6-inch slot is cut to pass the conduit through the ceiling. A keyhole saw is used to cut the slot. A hanger is installed as shown, to support the fixture. This hanger is nailed to the joists before the conduit is attached to the box. Note that whenever boxes are installed in walls or ceilings, it is essential to mount the box so that it will be flush with the surface of the plaster or building board in the completed construction. Wires can be pushed through short runs of conduit, but must be "fished" through long runs.

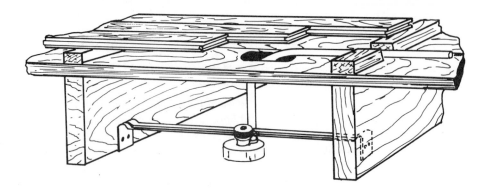

Figure 7-17 Thin-wall conduit run to a ceiling outlet

7–4 EXPOSED WIRING

Exposed wiring is installed chiefly in industrial wiring systems, and is occasionally utilized in old residential work. A run of exposed wiring is protected by a surface metal raceway. Figure 7-18 illustrates various types of metal raceways and fittings. A raceway can be compared to thin-wall conduit in some respects, in that the installation is a continuous assembly electrically and mechanically. However, a raceway is never bent and fittings are utilized

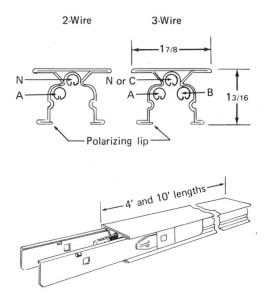

Figure 7-18 Typical surface metal raceways and fittings

wherever the direction of run changes. Metal straps are employed to secure a raceway to the surface of a wall or ceiling. Wires are "fished" through the raceway after it is installed. Exposed wiring installation will be explained in greater detail later.

7–5 PUBLIC UTILITY REQUIREMENTS FOR UNDERGROUND SERVICE ENTRANCES

Before a service can be installed, the local public-utility office must be provided with an accurate load information (current demand) report and the date on which service will be required. In areas where local ordinances require permits and inspection, these must be obtained and the wiring of the residence approved before the public utility can provide service. A residence can be served with an underground service if the lot is located in an area that is supplied from an existing underground distribution system. Note that the underground service lateral is installed, owned, and maintained by the public utility from the utility's distribution line to the termination

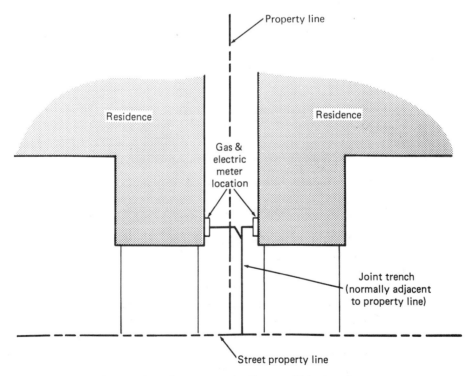

Figure 7-19 Location of meters, service lines, and joint trench

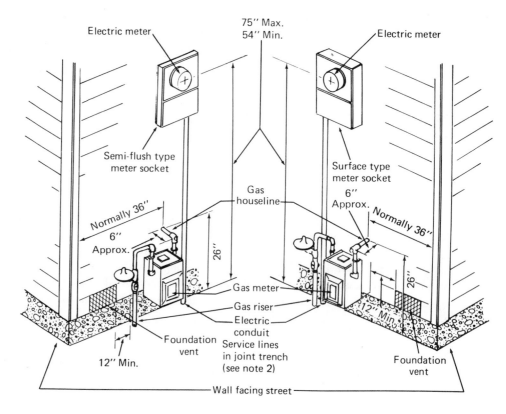

Figure 7-20 Residential service installations showing general requirements

Notes:
1. If in separate enclosures, meters may be recessed. Consult the company for requirements.
2. Meters may be offset horizontally by a maximum
of 36 inches.
3. Meters must be adequately protected if located
in areas where damage could occur.
4. See Fig. 7-22 for material requirements.

facility of the residence, usually a watt-hour meter as shown in Figure 7-19.
General arrangements and material requirements are shown in Figures 7-20
and 7-21.

A property owner who requests underground service is required to provide
or make arrangements and pay for the trenching and backfill in accordance
with utility specifications. If the total length of the service line is 100 feet or
less, the service conductors are provided at utility expense. On the other
hand, if the service line exceeds 100 feet in length, the property owner is
charged with material costs for conductors in excess of 100 feet. If a residence
is located in an area served from an overhead system, underground service
may be provided from an underground riser installed on an existing pole.

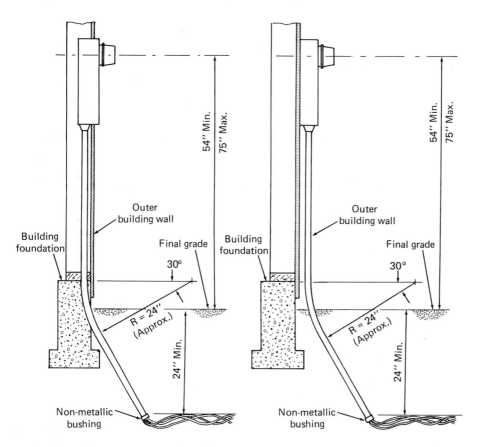

Recessed Mounted Service Termination Enclosure   Surface Mounted Service Termination Enclosure

MATERIAL

ITEM	DESCRIPTION
1	Service Termination Enclosure (Combination Meter Socket Panel illustrated)
2 *	Meter, Watthour, Single Phase, Socket Type
3	Conduit, 1¼″ Min. I. D. for No. 2 Al Service Cable, Material per Code Reqm't. (see note)
4 *	Service Cable, 600 V., Direct Buried, No. 2 Al Min.

* items to be supplied by utility.

* Note:
Conduit on or within the customer's building shall be in accordance with applicable building codes except that aluminum conduit shall not be installed below ground.

Figure 7-21 Typical underground service attachment material requirements

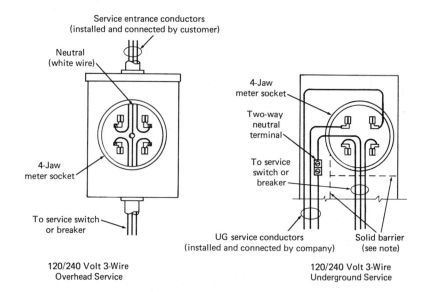

120/240 Volt 3-Wire
Overhead Service

120/240 Volt 3-Wire
Underground Service

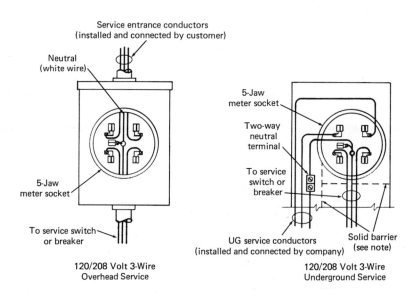

120/208 Volt 3-Wire
Overhead Service

120/208 Volt 3-Wire
Underground Service

Note:
    Combination switch and meter sockets shall include a solid barrier
between sections containing metered and unmetered wires. Openings
in any enclosure containing unmetered wires shall be sealable.

Figure 7-22 Connections for a meter socket

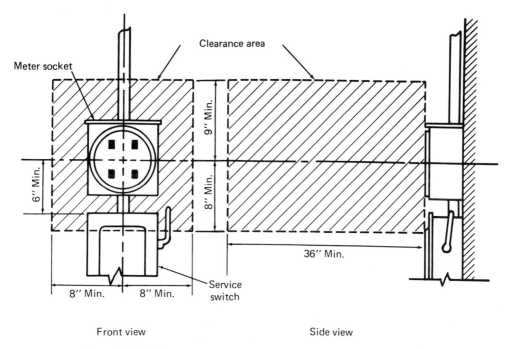

Figure 7-23 Meter socket clearance requirements

In this case, in addition to the foregoing requirements, the property owner is required to pay the material costs of the pole riser facility, of any required conduit in the public right of way, and of any conductor in excess of 100 feet as measured from the top of the riser.

A property owner is also required to furnish and completely wire a four-terminal socket (five-terminal for 120/208-volt service) without circuit-closing devices, as shown in Figure 7-22. If the ampacity of the main service switch does not exceed 125 amperes, any U-L approved socket may be used, provided it has (a) beryllium copper alloy (or equivalent) jaws, (b) approved terminals for use with aluminum as well as copper conductors, and (c) for underground service an insulated two-way neutral terminal lug for connecting the aluminum service cable neutral wire to the neutral conductor of the residential wiring system. If the ampacity of the residential service switch or breaker is 126 to 200 amperes, only 200-ampere sockets approved by the utility may be utilized. Again, if the current demand is in excess of 200 amperes, current transformers are required, as specified by the public utility.

A meter socket must be mounted so that the socket jaws are vertical and plumb, regardless of whether the conduit or cable enters the socket vertically or horizontally. Note that the meter socket must always be located on the source side of the service switch. Observe also that the sequence of equipment is (a) service entrance wiring, (b) meter, (c) switch, and (d) fuse or breaker. A main service switch or breaker must be installed at a readily accessible

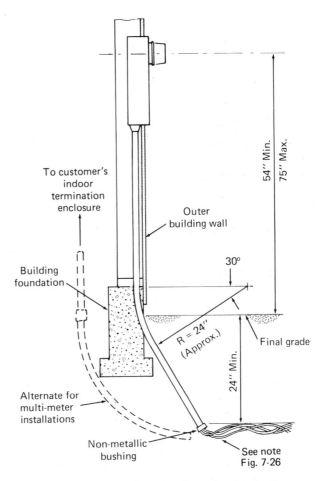

To customer's
indoor
termination
enclosure

Outer
building wall

Building
foundation

54" Min.
75" Max.

30°

R = 24"
(Approx.)

Final grade

24" Min.

Alternate for
multi-meter
installations

Non-metallic
bushing

See note
Fig. 7-26

Figure 7-24 Recessed mounted service termination enclosure

location near the entrance of the conductors either inside or outside the building wall. If the service switch is installed in a location exposed to the weather, it must be of weatherproof design or protected by a weatherproof enclosure. Suitable standing space (at least 3 feet) must be maintained in front of the meter to allow installation, testing, and reading. Eight inches of clear space is normally required below the center line of the socket, but this clearance may be reduced to 6 inches, as shown in Figure 7-23, if the obstruction is the service switch.

A recessed mounted service termination enclosure is shown in Figure 7-24. Note that the meter must be mounted at a height between 54 and 75 inches above the final grade (ground level). Figure 7-25 depicts a surface-mounted service termination enclosure. In this example, the conduit does not pass through the foundation of the building. Note that, in general, public ultilities

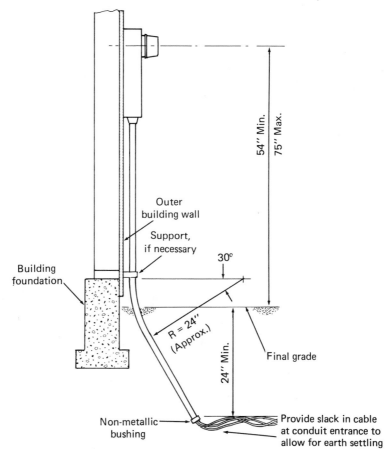

54" Min.
75" Max.

Outer
building wall

Support,
if necessary

30°

Building
foundation

R = 24"
(Approx.)

24" Min.

Final grade

Non-metallic
bushing

Provide slack in cable
at conduit entrance to
allow for earth settling

Figure 7-25 Surface-mounted service termination enclosure

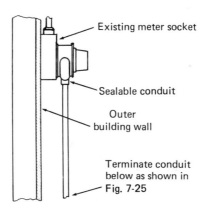

Existing meter socket

Sealable conduit

Outer
building wall

Terminate conduit
below as shown in
Fig. 7-25

Figure 7-26 Conversion from
overhead to underground service

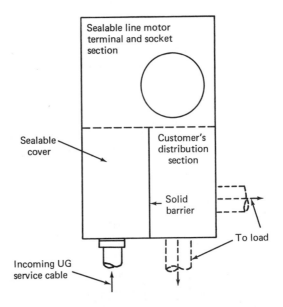

Figure 7-27 Typical service termination enclosure combination meter socket panel

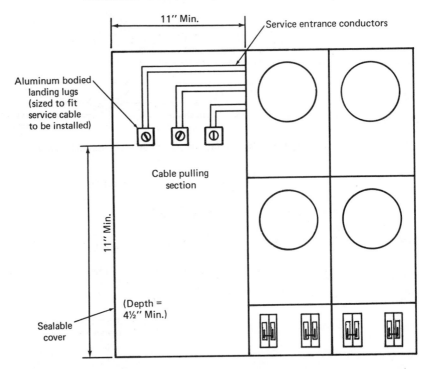

Figure 7-28 Typical combination service termination enclosure meter socket panel for multi-unit residential use

require 3-inch conduit from the meter to the underground service. Figure 7-26 depicts a typical service termination-enclosure combination meter-socket panel. In the case of multi-unit residential installations, a larger panel is employed with several meters, as shown in Figure 7-27. Multi-unit installations are explained in greater detail later.

STUDENT EXERCISE

This exercise provides familiarity and practical experience with conduit.

1. Obtain three sections of conduit approximately 5 feet in length and a conduit bender.

2. Use the conduit bender to make a right-angle bend in a section of conduit. Keep your foot on the conduit bender while applying leverage to the handle. Try to make the radius of the bend approximately eight times the inner diameter of the conduit as depicted in Figure 7-28. (This is the minimum radius permitted by the NEC.)

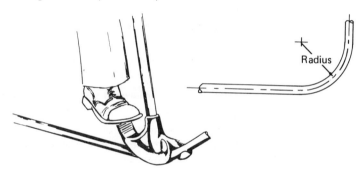

Figure 7-29 A right-angle, or ell bend

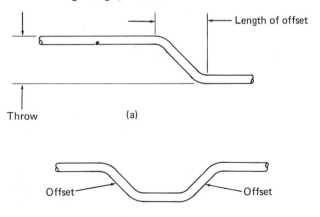

Figure 7-30 (a) An offset bend (b) a saddle bend

3.   Make an offset bend in another section of conduit. Try to form a throw of 6 inches and a length of offset of 10 inches. (See Figure 7-30(a).)

4.   Make a saddle bend in another section of conduit. Try to form a saddle 2 feet in length, with 6-inch offsets, as shown in Figure 7-30(b).

STUDENT EXERCISE

This exercise provides familiarity and practical experience with insulated switch and receptacle outlet boxes.

1.   Obtain a plastic outlet box, a short length of 2 × 4 stud, a hammer, and a short length of 3/8″ metal rod.

2.   Observe the construction of the box. Insulated boxes do not have cable clamps and are used only with nonmetallic sheathed cable.

3.   Note how the box is designed to be attached to a stud with nails, in much the same manner as a metal box.

4.   With the hammer and metal rod, try to punch out the knockouts cleanly.

5.   Nail the box to the side of the 2 × 4 section of studding. Be careful not to crack the box.

6.   Observe the threaded holes provided in the edge of the box for mounting a toggle switch or receptacle outlet.

7.   If the threaded holes are covered with a thin film of plastic at the top, use the point of a knife to remove the film.

8.   Observe that the knockouts are sufficiently large to pass nonmetallic sheathed cable easily. In practice, the cable must be secured to the stud at a distance not greater than 8 inches from the knockout.

9.   Note that although an insulated box cannot be grounded, a receptacle outlet can be grounded by connecting a grounding wire to the green terminal on the outlet strip.

STUDENT EXERCISE

This exercise provides familiarity and practical experience with basic ground clamps and straps.

Obtain the two types of conduit ground clamps and a ground strap, as illustrated in Figure 7-30. Also obtain a short section of ordinary water pipe, and a short length of No. 12 bare copper wire.

1.   Inspect the conduit ground clamps, and observe the provisions for securing each clamp tightly around a water pipe. Also observe the terminal facilities for connecting a ground wire to the clamp.

2.   Slide the clamps over the water pipe, and tighten them securely in place.

3. Insert a copper wire through the threaded conduit opening in each clamp, and connect the end of the wire to the clamp.

4. Inspect the ground strap, and observe how it is tightened in place after being wrapped around the water pipe.

5. Mount the ground strap on the pipe, and tighten it securely.

6. Connect the end of a copper wire to the ground strap, and tighten the terminal nut.

7. Approximately what diameters of water pipe does each of the ground clamps accommodate?

8. What diameter of conduit does each of the ground clamps accommodate?

9. Approximately what diameters of water pipe does the ground strap accommodate?

10. Would it be good practice to scrape or sandpaper the surface of the water pipe which is contacted by the ground clamp or strap?

QUESTIONS

1. In what situations may flexible armored cable be installed?

2. What type of switch boxes must be used with armored cable?

3. Why is Greenfield or flexible conduit easier to install than rigid conduit?

4. What is a hickey?

5. How would you make an ell bend? A saddle bend? An offset bend?

6. What is a snake utilized by an electrician in pulling wires?

7. What is wiring inside of walls called?

8. How is nonmetallic cable run when it crosses studs?

9. How much insulated wire should be left to permit connections at the end of conduit runs?

10. At what intervals should armored cable be strapped?

11. What is the purpose of the fiber bushing that is inserted in the end of an armored cable?

12. How is a knockout generally removed from a box?

13. How is a hole closed if a knockout is accidentally punched out of a box?

14. How is a grounding wire usually connected to a box?

15. Why is EMT easier to install than rigid conduit?

16. What is the maximum number of 90° bends allowed between boxes?

17. What is the approximate distance that convenience outlets are usually mounted above floor level?

8 WIRING INSTALLATIONS

8–1 MOUNTING OUTLETS AND SWITCHES

A square switch and receptacle box with a side-mounting bracket is pictured in Figure 8-1(a). This type of bracket has inverted (protruding) points that help secure the box in position on the stud. In addition, nails are driven into the stud through the holes in the bracket. In some cases, screws may be used instead of nails. Figure 8-1(b) shows the placement of a standard switch box with side-mounting bracket against a stud. As seen in Figure 8-1(c), a steel box support is used for mounting a receptacle box between studs. It is also used to mount one or more switch boxes in any position. Note that a square box serves as a junction box for splices, if provided with a plain cover. Figure 8-1(d) depicts a hanger mounted between tie beams for support of a ceiling outlet.

Figure 8-2 shows a typical installation of two convenience outlets, a switch box, and a ceiling outlet, interconnected with wires protected by conduit. Note that the conduit is anchored by notching the studs. Where conduit runs up the side of a stud or joist, it should be secured at 6- or 8-foot intervals with pipe straps or clamps, as depicted in Figure 8-3. Note that stud-notching can be avoided by running conduit across subfloors, as shown in Figure 8-4. Notches are made in the plate to accommodate the conduit at each bend upward to an outlet. Conduit may be laid directly on the rough floor. Then furring strips or sleepers are installed up to the conduit and are continued on the other side of the conduit up to the rough wall. Thus, the finished floor covers the conduit when it is nailed to the furring strips.

When switch or outlet boxes are installed in old-building wiring systems, the boxes must be mounted between studs, customarily at a distance of 4 or 5 inches from the nearest stud. Note that switches are usually mounted

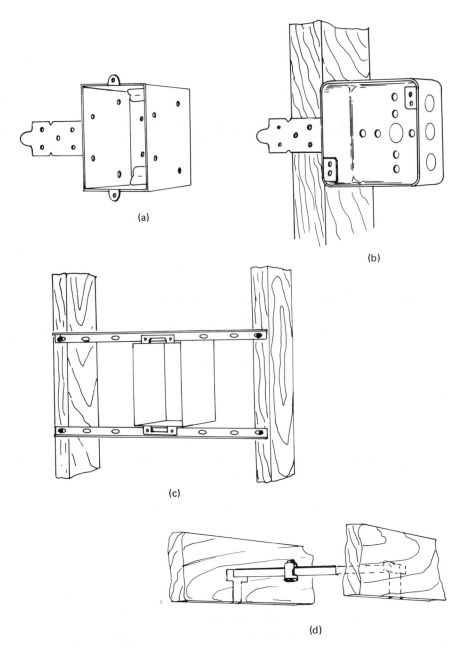

Figure 8-1 (a) Square switch and receptacle box with side-mounting bracket positioned on stud (b) placement of a standard switch box with side-mounting bracket (c) steel box support for mounting a receptacle box (d) hanger for supporting a ceiling outlet

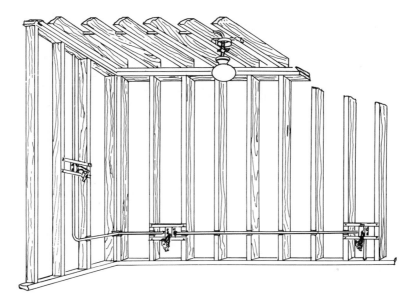

Figure 8-2 Typical installation of convenience-, switch-, and ceiling-outlet boxes

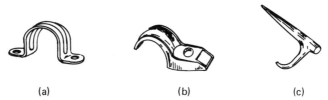

(a)   (b)   (c)

Figure 8-3 (a) Two-hole conduit strap (b) one-hole strap (c) nail strap

from 48 to 54 inches above the floor. Wall fixture outlets are generally mounted from 66 to 70 inches above the floor. Switches are located at the opening side of doors—not at the hinged side. Figure 8-5 shows the procedure to be followed for installing a box in old work. Drill four 1/2-inch holes, as shown, to permit entry of a hacksaw blade. Mount the blade so that the saw cut is made by drawing the blade outward. If the electrician holds his hand against the plaster, it will assist in preventing cracking. Armored cable or nonmetallic sheathed cable is used in this type of installation.

If a box is to be installed in building board instead of lath, a box with expansion anchors is utilized, as shown in Figure 8-6. After the cable has been secured in the box, push the box into the opening so that the front brackets are against the wall. Then tighten the side screws on the box until

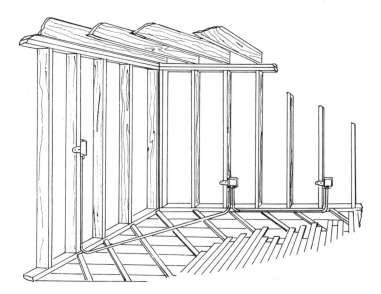

Figure 8-4 Example of conduit run across a sub-floor

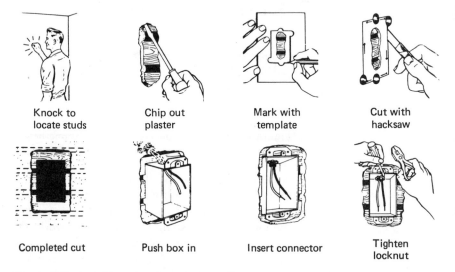

Knock to locate studs

Chip out plaster

Mark with template

Cut with hacksaw

Completed cut

Push box in

Insert connector

Tighten locknut

Figure 8-5 Installing a box in a plaster wall

the side brackets are pressing snugly against the inner surface of the wall. Wiring is usually installed with boxes 2-1/2 inches deep. However, if there is less than 2-1/2 inches of clearance, a shallow box may be used. Any box less than 1-1/2 inches deep is called a shallow box. Wiring procedures are made easier by employing large boxes. Note that two single boxes can be combined or ganged, as pictured in Figure 8-7, to double the size.

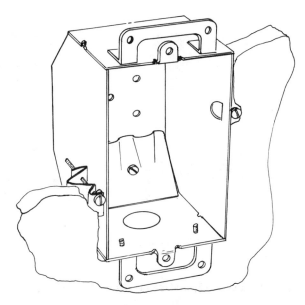

Figure 8-6 A box with expansion anchors

Ceiling boxes are easily installed from above, provided that there is access. However, when space above is inaccessible, installation must be made from below, as depicted in Figure 8-8. Plaster is chipped away to the size of a shallow box and the center lath is cut away with a hacksaw blade. A hanger is then inserted as shown, the locknut is removed, and the wire is placed through the threaded stud. If the stud is held above the ceiling with one hand, and the wire is pulled with the other hand, the hanger will center itself. Then the wire from the hanger is pulled through the center knockout and the locknut is turned up the threaded stud. Note that it is usually impractical to use conduit in old work. Details of installing armored cable and NMC are explained in a following chapter.

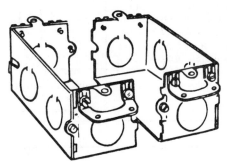

Figure 8-7 Two boxes combined, or ganged, together

Figure 8-8 Installation of a ceiling box

8–2 LIMITATIONS ON NUMBER OF WIRES IN OUTLET BOXES

Each size of box will accommodate a specified number of wires. In case more wires happen to be needed in a circuiting situation, a suitable layout must be planned so that no box contains an excessive number of wires. Figure 8-9 gives a convenient summary of these requirements for the most widely used boxes with copper or aluminum wires. Note that a bare grounding wire may need to be included in the wire count. Aluminum wire is easier to work with than copper because it is not as springy. In turn, an

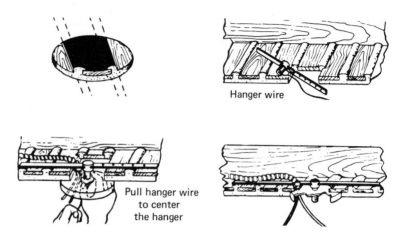

National Electrical Code specifies aluminum wire one size larger than copper. This means fewer aluminum wires in a box. (Check local code to see if bare grounding wires must be included in count.) Numbers given here are for boxes without internal cable clamps, studs, or hickeys (threaded fixture connectors). Deduct one wire for each of these included in box.	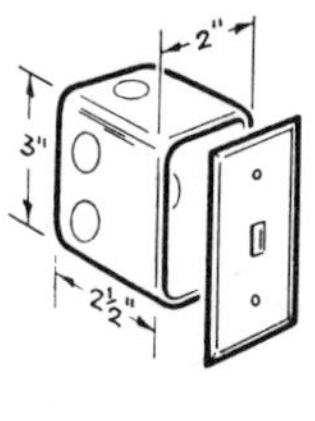 Rectangular with flush wiring device such as switch or receptacle	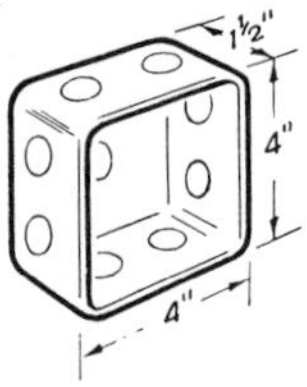 Octagonal with no flush wiring device	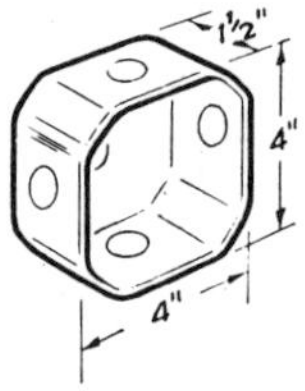 Square with no flush wiring device
Copper wire #12	5	7	9
Aluminum wire #10	4	6	7

Figure 8-9 Number of wires permitted in various boxes

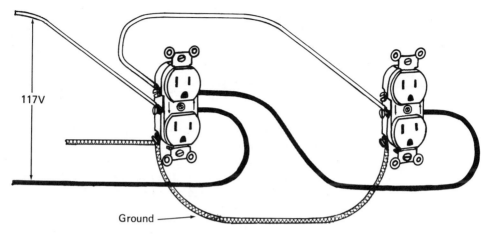

Figure 8-10 Most receptacles have two terminal screws on either side

outlet installed with aluminum wire is easier to assemble and to hold in position while the screws are being tightened. Most receptacle boxes have two terminals on each side, as pictured in Figure 8-10, to allow convenient placement of wires in boxes.

8–3 INSULATED BOXES

Nonmetallic boxes are being used more, for reasons of economy, than in the past. These boxes are made of an insulating material such as porcelain or Bakelite.® Installation of nonmetallic boxes avoids possibile trouble in case a hot wire should accidentally make contact with the inside surface of the box. However, more care is required when installing nonmetallic boxes, because they are not as rugged as metallic boxes and they will be damaged if excessive force is applied. Note that insulated boxes cannot be used with flexible armored cable. Indoor-type plastic-sheathed cable or dual-purpose plastic-sheathed cable is generally used with insulated boxes. It is good practice to always install insulated boxes in damp locations, such as barns in rural wiring systems, not only from the standpoint of safety, but also because an insulated box does not rust and will last much longer than a metallic box.

8–4 OLD WORK PROCEDURES

Old work (modernization of wiring systems) often calls for ingenuity as well as experience. Lath-and-plaster walls and ceilings may be found instead of sheetrock or building board. Careful work is required with plaster and wallpaper when outlets are installed, or damage may occur. In comparatively old houses, the obsolete knob-and-tube wiring system may have been used. Wire sizes in old houses are often too small to supply additional loads, and may be found inadequate for the loads that are already in use. Therefore, the electrician must review the existing wiring system, and install additional circuits from a new add-on panel, if required. Code requirements are less strict for old-work projects, but principles of good practice should be followed, nevertheless.

Regardless of the existing wiring system, plastic-sheathed cable, dual-purpose plastic-sheathed cable, flexible armored cable, or Greenfield is ordinarily used in most new circuiting that is installed in old work. In other words, it is usually impractical to install rigid conduit or EMT in a finished building. When it is impossible to work behind walls, as when lath and furring are attached to a brick wall, surface metal raceways may be used, as shown in Figure 8-11. Raceways are similar to conduit, and the wires are pulled through after the joints have been completed and the outlets installed. It is preferable to install raceways in corners between the molding and wall, and between the door jamb and wall, where it is least noticeable. Raceways are always grounded, just as conduit and armored cable are grounded.

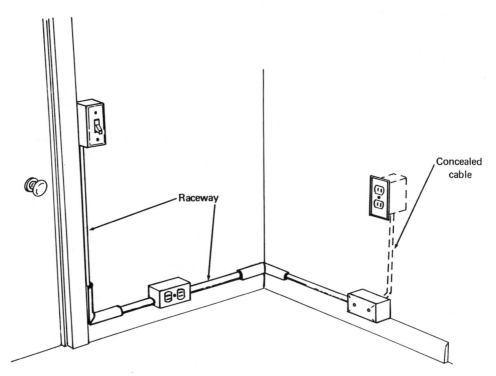

Figure 8-11   Surface metal raceways are sometimes required in old work

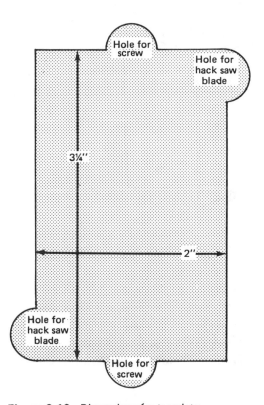

Figure 8-12   Dimensions for template

Mounting and installation of boxes was explained previously. If necessary, after the job is completed, any cracks or holes can be filled with spack-
ling or plaster of Paris. In case comparatively large areas are accidentally
damaged, they can also be patched with plaster of Paris. In addition, the
patch may need to be tinted to match the original plaster. Note that cutouts
should be made with a template. A template can be cut from a sheet of stiff
cardboard or plastic, as indicated in Figure 8-12. The electrician uses a pencil
to trace around the inside edges of the template, thereby outlining the
required opening accurately on the plaster. Note that in very old houses,
the plaster may be unusually brittle and easily damaged. It is good practice
in this situation to hold a strip of wood against the plaster (instead of the
hand) along the outer edge of the cutting line, or, the strip of wood may be
temporarily screwed to the wall. After the opening is completed, the screw
holes are patched with plaster of Paris and tinted as required.

Note that all holes, open spaces, chipped areas, or gaps must be patched
in order for the finished job to pass inspection, according to the NEC. Old
houses sometimes have plaster walls that are papered. In such a case, the

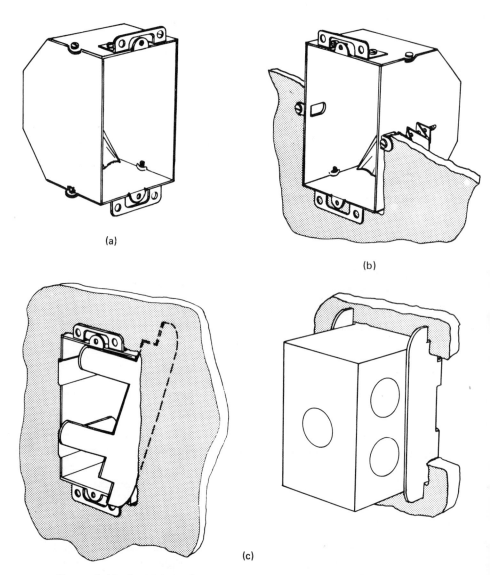

(a)

(b)

(c)

Figure 8-13 Special old-work boxes for plaster or sheetrock walls (a) beveled corner box (b) the Grip-Tite® box (c) installation of metal box supports

paper is cut away with the plaster and lath when the opening is sawed out. If difficulty is encountered in fishing the cable, one or more additional openings may need to be cut through the wall in order to pull or push the cable into place. In this process, the wallpaper must be properly loosened, cut, and finally replaced. Wallpaper paste can be softened in the work area by applying a damp cloth against the paper. Then a three-sided flap can be cut in the paper with a razor blade. This permits the flap to be folded up, so that a hole can be cut through the plaster and lath. Finally, after the cable has been installed, the hole must be repaired and plastered, and the flap pasted back in place.

Various special types of boxes are available for old work. Previous mention was made of expansion-type boxes. Beveled-corner boxes, shown in Figure 8-13(a), are easier to install in old work and do not utilize connectors. Clamps are provided for NMC or old-style loom sheathing. This type of box is also available with knockouts to accommodate connectors. In Figure 8-13(b), note the Grip-Tite clamps at the sides of the box. After the box is seated in place, the side screws are tightened, which expands the clamps and brings them up snugly against the wall behind the box. This type of box is also very useful for installation in sheetrock walls. Conventional boxes can be secured in place to good advantage by inserting additional metal box supports on each side, as depicted in Figure 8-13(c). This box support is worked in beside the box so that its ends rest against the wall behind the box. Then the support is drawn forward snugly and the two tabs are bent down over the edge and inside the back of the box.

Large-size boxes 2-1/2 inches deep are preferred, as they are easier to work with. However, small and/or shallow boxes must sometimes be used. Note that a 4-inch square box can be fitted with a suitable cover, as shown in Figure 8-14, and used as a switch box. Figure 8-15 shows the cleat-type box that is somewhat similar to a "Grip-Tite" box. A cleat-type box has a pair of folding side cleats that are unfolded and drawn up against the back of the wall by tightening a pair of screws. Boxes as shallow as 1/2 inch are permitted

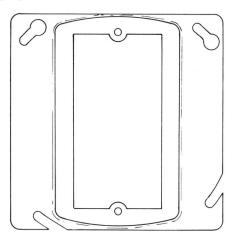

Figure 8-14 A cover used with a four-inch square box for use as a switch box

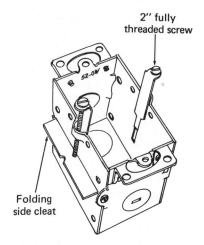

Figure 8-15 Cleat type of old-work box

Figure 8-16 Rotary hack-saw

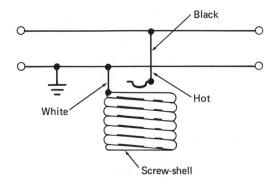

Figure 8-17 Fixture circuit connected in accordance with the National Electrical Code

by the NEC for installation in difficult situations. However, where boxes cannot be installed in walls, surface wiring must be utilized, as explained previously. When ceiling boxes are installed, a rotary hacksaw may be used to cut a hole with the diameter of a shallow fixture box. In other words, the rotary hacksaw is used instead of chipping the plaster away with a chisel or screwdriver. A rotary hacksaw is illustrated in Figure 8-16.

As noted previously, all present-day fixtures have a black wire and a white wire. In turn, the socket is connected as shown in Figure 8-17. This mode of connection ensures that the screw-shell will be "dead" when there is no lamp in the socket, thereby reducing shock hazard. When repairing fixtures in old work, the electrician may encounter circuits in which the screw-shell of a fixture has been connected to the "hot" side of the line. In such a case, it is good practice to reverse the fixture connections in accordance with the requirements of the NEC.

8–5 INSTALLATION OF OUTLET CABLES

Installation of outlet cables in old work requires various procedures, depending on the construction of the building. Figure 8-18 shows how armored cable can be fished through walls, utilizing a hole bored through from the cellar. Fish tapes, or ordinary iron wire with a hook bent on the

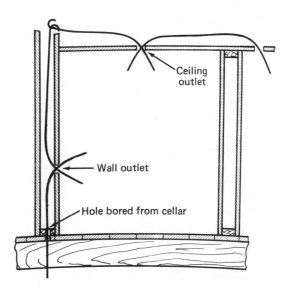

Figure 8-18 Cable can be pushed through a hole bored from the cellar and fished through the walls

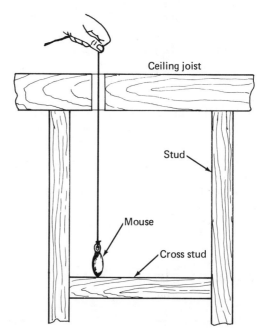

Figure 8-19 Locating cross studs with a mouse

end, may be used to draw the cable through the walls. A mouse, or small plumb bob, is very useful to check for cross studs, as depicted in Figure 8-19. To bypass a cross stud, the electrician bores two holes through the wall, as shown in Figure 8-20. These holes are drilled at 45° angles to the cross studs, so that the cable can be passed around the stud. In some cases, the plaster must be repaired after the cable has been fished through.

Figure 8-21 shows an example of cable installation from one outlet to another through a wall and underneath a floor. If either of the outlets is installed in an interior wall, it is often possible to drill upward to the partition between the walls, as was shown in Figure 8-18. In the basement, a cable strap should be placed every 3 feet to secure the cable to sides of beams or joists. To install a cable from a wall switch to a baseboard outlet, remove the baseboard and cut holes as indicated at (1) and (2) in Figure 8-22. Then notch a channel in the plaster to pass the cable. Attach a connector to the cable and fish the cable from (1) to the hole for the wall-switch outlet. Finally, run the cable through hole (2) to the hole for the baseboard outlet. Replace the baseboard, connect the cable, and finish installing the boxes. To install a cable from one baseboard outlet to another, the same basic procedure is followed, as shown in Figure 8-23.

When installing outlets back-to-back in adjacent rooms, as shown in Figure 8-24, it is advisable to join the outlets with conduit or with a threaded nipple and locknuts. If cable is installed from an outlet to a fixture in the room below, a diagonal hole is bored, as seen in Figure 8-25. A fish tape is

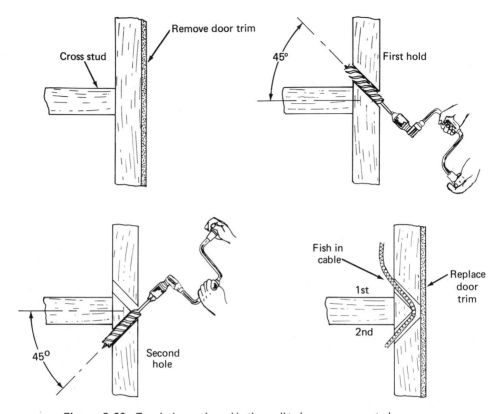

Figure 8-20 Two holes are bored in the wall to bypass a cross stud

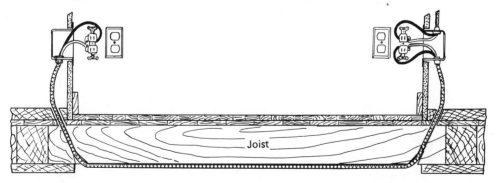

Figure 8-21 Left outlet connected to right outlet with cable running through floor and across the basement

pushed up through the hole in order to pull the cable through and up behind the wall to the outlet. When a cable or wires must be fished for a considerable distance, the conductors must be firmly secured to the fish tape, as shown in Figure 8-26. Otherwise, the hook is likely to be pulled open and the job must then be started all over again.

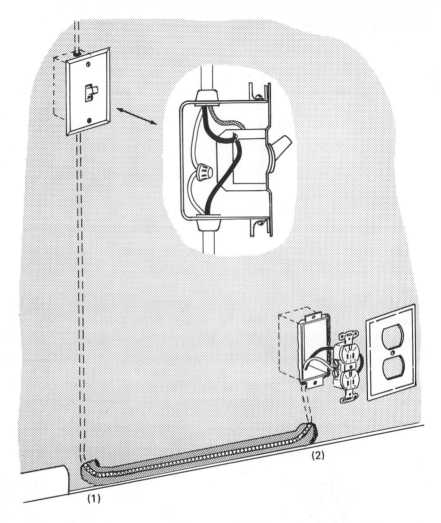

Figure 8-22 Running a cable from a wall switch to a baseboard outlet

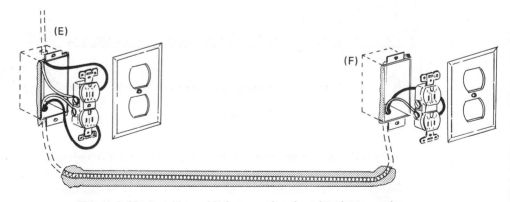

Figure 8-23 Installing cable from one baseboard outlet to another

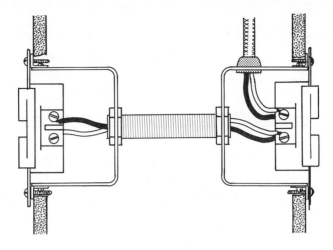

Figure 8-24 Installing back-to-back outlets in adjacent rooms

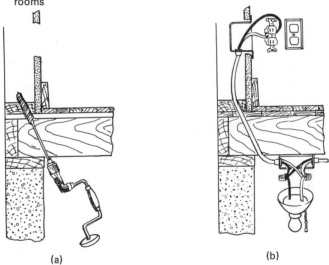

(a)

(b)

Figure 8-25 Installing cable from an outlet to the fixture in the room below (a) drilling hole from basement (b) cable pulled through

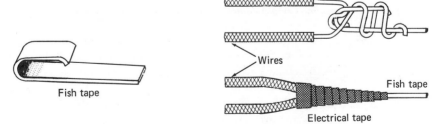

Fish tape

Wires

Fish tape

Electrical tape

Figure 8-26 How to secure wires firmly into a fish tape hook

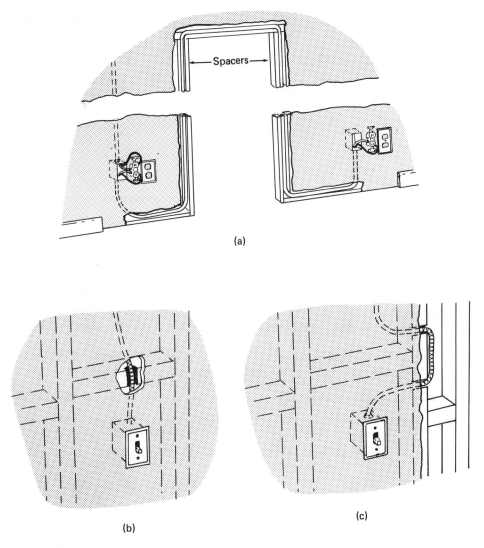

Figure 8-27 Installing cable around a door frame (a) notching of spacers and wall (b) passing cable at center of header (c) passing cable at end of header

It is often necessary to install cable around a door frame from one outlet to another. This is done as depicted in Figure 8-27. First, the door trim and the baseboard are removed. Then the wall and the spacers between the door frame and jamb are notched to pass the cable. If there are headers to be passed, they are also notched. After the cable is installed, the trim and baseboard are replaced, and the wall is patched with plaster of Paris and tinted if required.

Figure 8-28 shows the details of fishing a wire. In this example, two diagonal holes are drilled, (a) a hole is cut in the wall, and another hole is cut in the ceiling. A horizontal hole is drilled to join the two diagonal holes (b). Then the electrician pushes a 12-foot fish wire, with hooks on each end, through the hole on the second floor. One end of the wire is then pulled out from behind the wall and through the opening for the switch outlet on the first floor (c). Next, a 25-foot fish wire, with hooks on each end, is pushed through the ceiling outlet, as shown by the arrows in the diagram (d). This wire is worked back and forth until it hooks the first wire. Either one of the fish wires may then be withdrawn to slide the hooks together (e). Finally, the shorter wire is pulled through the opening for the switch outlet. After the shorter fish wire is removed, the cable is attached to the long fish wire for drawing through the wall and ceiling spaces (f).

When cable is installed across joists or beams, attic floor boards must be lifted, as shown in Figure 8-29, so that joists can be notched and holes bored through obstructions. Boards can be lifted without undue difficulty if they do not have tongues and grooves. Small holes are drilled at the ends of a section to be lifted, and the board is cut with a thin keyhole saw. In turn, the end of the section can be pried up and lifted out. After the cable is

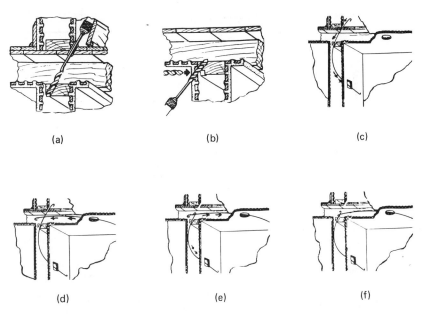

(a)   (b)   (c)

(d)   (e)   (f)

Figure 8-28 Typical cable fishing procedure (a) boring the first diagonal hole (b) boring the second diagonal hole and the horizontal hole (c) insertion of the shorter fish wire (d) insertion of the longer fish wire (e) bringing the hooks together (f) shorter wire is pulled to bring long wire through outlet opening

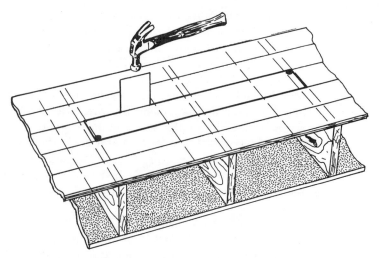

Figure 8-29 Tongues are cut from floor boards with a painter's scraper and hammer

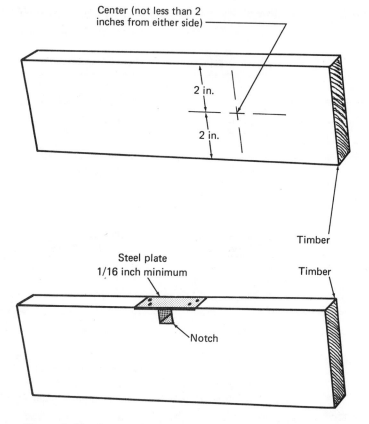

Center (not less than 2 inches from either side)

2 in.

2 in.

Timber

Steel plate 1/16 inch minimum

Timber

Notch

Figure 8-30 Approved methods of drilling and notching timbers

installed, the board is replaced, with cleats tapped in as shown in the figure, so that the board has a firm footing. Finally, the drill holes are filled with wooden plugs. When tongue-and-groove flooring is to be lifted, it is necessary to first cut the tongues free. This is done with a painter's scraper and a hammer. Some electricians prefer to start the job with the scraper and then to do most of the cutting with an extra-thin hacksaw blade. This method has the advantage of cutting through any stray nails without difficulty. It is good practice to lift boards over at least three joists, so that the replaced board has a firmer footing.

All building codes must be observed when passing cables through studs, joists, or rafters. In addition, the NEC requires that the holes must be drilled as near the center of the timber as possible, and that holes must be at least 2 inches from an edge. This requirement is depicted in Figure 8-30. Notches must be covered by steel plates not less than 1/16 inch thick, to protect the cable from nails. Note also that if ceilings are constructed of combustible material, a disk of asbestos or other fire-resistant substance should be placed between the ceiling and the canopy (top) of a ceiling fixture. Figure 8-31 shows how a fireproof disk is inserted when it is required.

Methods of mounting large drop fixtures are shown in Figure 8-32. In (a), the hanger support is screwed on the threaded stud in the outlet box. Then the wires are connected, and the canopy is raised and anchored in position by a locknut. In (b), machine screws are inserted in the threaded holes of the metal strap. Then the center hole in the strap is slipped over the stud in the outlet box and secured in position by a locknut. Next, the wires are connected and the canopy is slipped into place over the machine

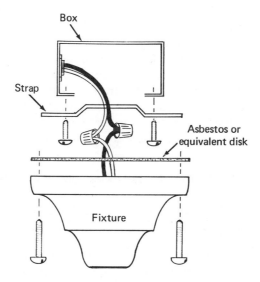

Figure 8-31 Insertion of fireproof disk in fixture canopy

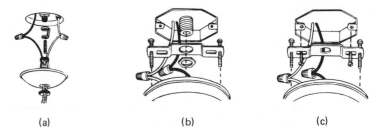

(a)   (b)   (c)

Figure 8-32 Mounting large drop fixtures (a) canopy secured by threaded stud (b) canopy secured by machine screws (c) strap secured to outlet box by machine screws

Figure 8-33 Installation
of a wall fixture

screws. Finally, the canopy is secured by a pair of cap nuts on the screws. In (c), an outlet box has been installed which does not have a threaded stud. Therefore, the strap is fastened to the ears of the outlet box with a pair of machine screws. Links in the chain can be removed or added as required to adjust the height of the fixture.

Wall fixtures are usually installed as depicted in Figure 8-33. A strap is secured to the ears of the outlet box by machine screws. A threaded nipple is then screwed into the strap for a suitable depth and the fixture is secured to the nipple by means of a cap nut. Many minor variations in fixture design will be encountered by the electrician. However, the installation procedure will become obvious after a careful inspection of the fixture construction.

8–6 INSTALLATION OF A THREE-GANG SWITCH BOX

Previous mention was made of ganged switch boxes. These are installed as illustrated in Figure 8-34. Procedure is as follows:

1.   Nail the ganged switch box mounting bracket to the stud, with the front edge protruding sufficiently to be flush with the front surface of the panelboard when the wall is finished.

Figure 8-34 Mounting a three-gang switch box (Courtesy Leviton, Inc.)

2. Insert cables through knockouts in the boxes and secure the cables with the cable clamps.

3. Make a cutout in the panelboard to pass the boxes, and nail the panelboard to the studding.

4. Install the switches, and connect the wires to the switch terminals.

5. Install the wallplate and secure by means of the machine screws that are provided.

8–7 BASIC TOOLS

Certain tools have been mentioned previously, and every electrician is supposed to have his own set of tools. Figure 1-19 shows a basic set of tools. The chief use of each tool is noted on the page to the left of that figure. A journeyman electrician will have various power tools available in addition to hand tools. For example, the hand brace will be supplemented by a power drill. An extension rod is often used with a power drill to drill holes overhead without the necessity for climbing up and down ladders. A keyhole saw is often supplemented by a saber saw. Carbide-tipped masonry drill bits are used in power drills to make holes in cement or brick walls. Rotary hacksaws have been noted previously. A magnetic nail indicator is often useful, not only to locate studs behind a wall, but also to detect nails that might damage a wood bit. A ship's auger is often used because it resists nail damage better than conventional wood bits.

This exercise provides practical experience in mounting old-work boxes and the use of metal box supports.

1.   Obtain a Grip-Tite box as shown in Figure 8-35(a) and a conventional box with a pair of metal supports as shown in Figure 8-35(b), a piece of wallboard, an auger, and a keyhole saw.

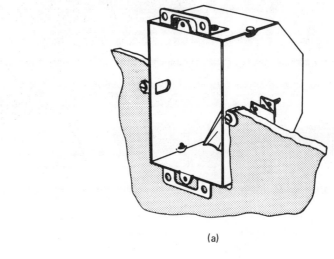

(a)

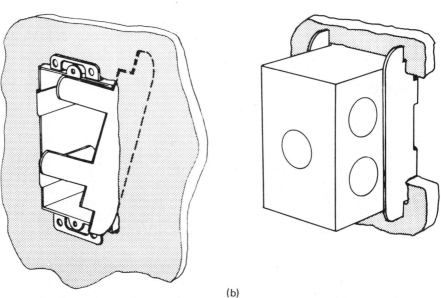

(b)

Figure 8-35 Mounting boxes in old work (a) Grip-Tite ® box (b) conventional box with supports

2. Mark the outline of the Grip-Tite box (exclusive of the front brackets) on the wallboard with pencil.

3. Use the auger and keyhole saw to make a cutout in the wallboard that will pass the box.

4. As shown in Figure 8-35(a), push the box into the cutout, with the front brackets resting against the wallboard.

5. Tighten the screws at the sides of the Grip-Tite box to bring the side brackets up snug against the wall.

6. Make another cutout in the wallboard to pass the conventional box.

7. As shown in Figure 8-35(b), push the box into the cutout, with the front brackets resting against the wallboard.

8. Insert a metal support on each side of the box. Work the supports up and down until they fit firmly against the inside surface of the wall. Then bend the projecting ears over the edges of the box and down so that the box is held securely in place.

STUDENT EXERCISE

This exercise provides practical experience in converting from two-wire to three-wire receptacle operation in old work.

1. Obtain a metal outlet box, a duplex receptacle outlet of the two-prong type, an outlet adapter, and a three-prong grounding plug.

2. Observe that the three-prong grounding plug will not fit into a two-prong receptacle outlet.

3. As shown in Figure 8-36, a three-prong grounding plug will fit into an outlet adapter, which in turn will fit into a two-prong receptacle outlet. Check out this arrangement.

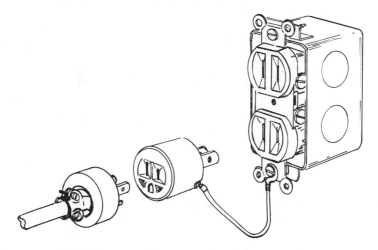

Figure 8-36 Converting from two-wire to three-wire receptacle operation

4. Observe that the outlet adapter has a grounding wire with a lug.

5. Connect the lug of the grounding wire under one of the screws that is used to mount the receptacle on the box, and tighten the mounting screws.

6. Note that this arrangement provides a ground connection to the grounding wire of the adapter, provided the box is grounded.

7. In case a box is not grounded, it is necessary to connect the grounding wire of the adapter to a ground point, such as a water pipe.

8. Could a ground wire be "fished" to a water-pipe location in a frame-construction building (in most cases)?

9. What type of building construction would require that a ground wire be run to a water pipe in a surface metal raceway?

QUESTIONS

1. How may a square box serve as a junction box?

2. At what intervals should conduit be secured to the side of a stud or joist?

3. How far are switches mounted above a floor?

4. How far are wall fixtures mounted above a floor?

5. What is the definition of a shallow box?

6. How many wires may be run into a standard outlet box?

7. May an insulated box be used with flexible armored cable?

8. Why might an add-on panel be required in an old-work project?

9. What is a Grip-Tite clamp?

10. How is a rotary hacksaw utilized?

11. What is the definition of a mouse?

12. When would a pair of outlets be joined with a threaded nipple?

13. Why must floor boards be lifted occasionally?

14. What is a canopy?

15. What is the purpose of a three-prong grounding plug?

9 BRANCH CIRCUITING

9–1 GENERAL CONSIDERATIONS

Three basic types of branch circuits are used in residential wiring systems. General-purpose branch circuits have an ampacity of either 15 or 20 amperes. They supply all lighting outlets throughout the house and all convenience outlets except those in the kitchen, dining room or dining areas of other rooms, breakfast room or nook, and laundry or laundry area. These general-purpose circuits are installed on the basis of one 20-ampere circuit for not more than each 500 square feet, or one 15-ampere circuit for not more than each 375 square feet of floor area. Outlets supplied by these circuits are divided equally among the circuits. Separate branch circuits should be provided for lighting and for convenience outlets in living rooms and bedrooms, and the branch circuits serving convenience outlets in these rooms should be equipped with split-wired receptacles.

Appliance branch circuits have an ampacity of 20 amperes and operate at either 117 or 234 volts. At least one 20-ampere branch circuit, equipped with split-wired receptacles, should be provided for all convenience outlets in the kitchen, dining room, or dining areas of other rooms, and breakfast room or nook. This circuit should also be extended to the laundry or laundry area to serve any convenience outlets not otherwise required to be served by individual equipment circuits. The use of three-wire circuits for supplying convenience outlets in the foregoing locations is an economical means for dividing the load and provides practical operating advantages. Such circuits supply greater capacity at individual outlet locations and lessen the voltage drop in the circuit. They also provide greater flexibility in the use of appliances. Note that for maximum effectiveness in utilization, the upper half of all receptacles should be connected to the same side of the circuit.

Table 9-1 Equipment and ampacity ratings requiring individual-equipment branch circuits

Item	Conductor Capacity
Range (Up to 21-kw rating)	50A-3W-115/230V
Combination washer-dryer or	40A-3W-115/230V
Automatic washer	20A-2W-115V
Electric clothes dryer	30A-3W-115/230V
Fuel-fired heating equipment (if installed)	15A or 20A-2W-115V
Dishwasher and waste disposer (if necessary	
plumbing is installed)	20A-3W-115/230V
Water heater (if installed)	Consult local utility

Table 9-2 Ampacity ratings of units for which individual-equipment branch circuits may be used

Item	Conductor Capacity
Attic fan	20A-2W-115V (switched)
Room air conditioners or	20A-2W-230V
central air-conditioning unit	40A-2W-230V
Food freezer	20A-2W-115 or 230V
Water pump (where used)	20A-2W-115 or 230V
Bathroom heater	20A-2W-115 or 230V
Work shop or bench	20A-3W-115/230V

Individual-equipment branch circuits are provided for the equipment tabulated in Table 9-1. Spare circuit equipment should be provided for at least two future 20-ampere, two-wire, 117-volt circuits in addition to those initially installed. If the branch circuit or distribution panel is installed on a finished wall, raceways are extended from the panel cabinet to the nearest accessible unfinished space for future use. Consideration should also be given to the provision of circuits for the commonly used household appliances and items of equipment tabulated in Table 9-2. Other equipment may need to be accommodated in particular installations. In some instances, one of the circuits recommended may serve two devices that are not likely to be used at the same time, such as an attic fan and a bathroom heater. Where a choice exists, fixed appliances should be operated on 234-volt circuits instead of 117-volt circuits.

In addition to the foregoing requirements, a number of different systems of electrical house heating might be specified. These systems are largely a matter of individual engineering design for the house under consideration and in keeping with local climatic conditions. Wiring and service capacity for such purposes must be considered for specific situations. Note that from 1 to 2 watts per cubic foot is required to heat the air in a room when the

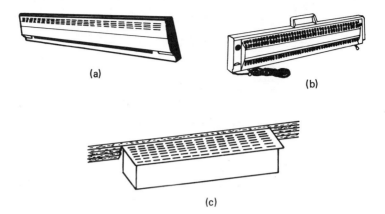

Figure 9-1 Typical baseboard electric heating units (a) permanent unit (b) portable unit (c) floor type (recessed) unit

temperature is near freezing outside. Baseboard heating units such as depicted in Figure 9-1 are commonly utilized. In addition, electric heaters may be built into walls or ceilings. Heat pumps are units that heat or cool and filter air, and control humidity. Heat pump current demands are basically the same as for comparable air conditioners. In a wiring system of fairly elaborate design, consideration should be given to the installation of branch-circuit protective equipment, served by appropriately heavy feeders, at locations throughout the house, rather than at a single location.

Note that all electric heating installations employ thermostats to automatically open or close the circuit when the temperature rises or falls past a preset limit. The location of a thermostat should be chosen to provide good control action. That is, the thermostat should not be mounted directly above a baseboard heating unit. It should be located away from the source of heat, and out of drafts that may arise through doorways, for example. A thermostat should be placed on a wall about 4 or 5 feet above floor level. Note that some thermostats are designed to switch a 117-volt circuit, while others are designed to switch low-voltage control circuits. Control circuits are generally used when fairly heavy currents must be switched.

Figure 9-2(a) and (b) depicts basic line-voltage and low-voltage thermostat circuits. Relays in low-voltage control circuits are ordinarily called *operators*. Line-voltage thermostat circuits are wired with conductors such as plastic-sheathed cable of suitable ampacity rating. Low-voltage thermostat circuits are wired with conductors similar to bell wire, as shown in Fig. 9-2(c). Concealed wiring is always utilized, unless it is impractical in an old-work situation. Note that baseboard electric heating units are available with built-in thermostats. This type of unit may be chosen when exposed thermostat wiring would otherwise be required.

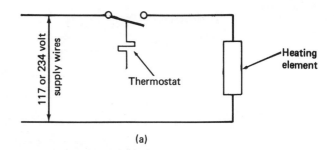

(a)

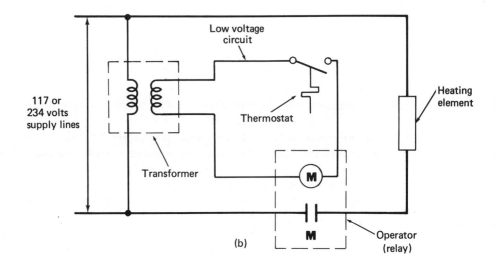

(b)

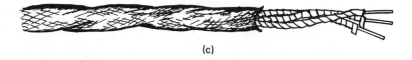

(c)

Figure 9-2 Basic thermostat circuitry (a) line-voltage type (b) low-voltage type (c) low-voltage thermostat cable

9–2 SPECIAL CIRCUIT REQUIREMENTS

After many wiring systems have been completed, it is discovered that they are less than might be desired because special circuit requirements were overlooked. For example, entrance signals should be provided at all regularly used entrances. Pushbuttons should be installed and connected to the door chime, with provisions for distinctive signals from different entrances. As noted previously, door bells and chimes operate from low-voltage circuits energized by step-down transformers. The transformer should be installed near the fuse box or circuit-breaker panel. Entrance-signal wires should be no smaller than No. 18 gauge.

In smaller residences, the door bell or chime is often installed in the kitchen if it will be heard throughout the house. If not, the unit should be installed at a more central location, which might be the entrance hall. In larger residences, extension signals are often necessary to ensure audibility throughout the living quarters. Extension signals may be provided in bedrooms, recreation rooms, or laundry rooms, for example. The door bell or chime may be supplemented by an intercommunicating telephone circuit, so that the caller may be spoken to without going to a door. Intercommunicating telephones should be operated from a power unit recommended and furnished by the manufacturer of the telephones and connected to the lighting service.

For the residence with accommodations for a live-in servant, additional signal and communication circuits should be considered. These include a dining-room-to-kitchen signal operated from a pushbutton attached to the underside of a dining table, or a foot-operated tread placed under the table. Another circuit may connect a flush (in-wall) annunciator in the kitchen with pushbutton stations in each bedroom, the living room, recreation room, porch, and so on. A flush intercommunicating telephone may be installed in the kitchen, with corresponding telephone stations in master bedroom, recreation room, and other desirable locations. Coaxial cable is preferred for wiring intercom circuits, as it provides quiet operation. The cable must not run closer than 2 inches to any AC line.

Automatic fire-alarm protection is recommended, with detectors in the basement oil-burner or furnace area and the storage space as a minimum. Alarm bell and test button should be installed in the master bedroom or other desirable location in the sleeping area. The operating supply is obtained from a separate bell-ringing transformer recommended and furnished by the manufacturer of the alarm equipment. Note that the alarm system should operate from a separate and independent transformer, rather than from the entrance-signal or other transformer, because of the increased possibility of failure due to causes originating in the entrance-signal or other system.

Telephone outlets also need consideration. At least one outlet should be installed, served by a raceway extending to unfinished portions of the basement or attic. The outlet should be centrally located and readily accessible from the living room, dining room, and kitchen. A desirable location for a second outlet is in the master bedroom. Where finished rooms are located on more than one floor, at least one telephone outlet should be provided on each such floor. The local telephone company should be consulted for details of service connection prior to construction, particularly in regard to the installation of protector cabinets and raceways leading to finished basements, where these are planned.

9–3 SPECIFICATION FORM

Table 9-3 shows a sample specification form that will serve as a guide in preparing wiring specifications for residences of small and medium size. Some additional provisions may be found necessary for very large houses, and can be included merely as extra articles. All outlets, the locations of wall switches, and the outlet or outlets controlled by each switch should be shown clearly on the floor plans that must be considered an essential part of the specifications.

Table 9-3 Sample specification form

SPECIFICATIONS FOR ELECTRIC WIRING IN THE DWELLING
TO BE ERECTED AT_____FOR_____

1. GENERAL—The installation of electric wiring and equipment shall conform with local regulations, the National Electrical Code, and the requirements of the local electric power supplier. All materials shall be new and shall be listed by Underwriters' Laboratories, Incorporated, as conforming to its standards in every case where such a standard has been established for the type of material in question.

2. GUARANTEE—The contractor shall leave his work in proper order and, without additional charge, replace any work or material which develops defects, except from ordinary wear and tear, within ___ months from the date of the final certificate of approval.

3. WIRING INSTRUCTIONS—Outlets, switches, and switch control shall be installed as shown on the plans. Contractor shall furnish switch and outlet bodies and plates and lampholders where indicated. Unless specifically contracted for, the hanging of lighting fixtures is not included.

4. WIRING METHODS—Interior wiring shall be

_____
            (Fill in wiring method)
No exposed wiring shall be installed, except in unfinished portions of basement, utility room, garage, attic, and other spaces that may be unfinished.

5. SERVICE ENTRANCE conductors shall be three No. _____ wires.

_____
            (Fill in wiring method)
6. SERVICE EQUIPMENT shall consist of_____
_____

7. DISTRIBUTION PANEL(S) — Branch-circuit equipment shall be_____
_____
            (Fill in type of equipment)
Distribution panel shall provide for termination of all initial branch circuits, plus two future branch circuits.

8. GENERAL-PURPOSE BRANCH CIRCUITS — At least _____ general-purpose branch
            (Number)
circuits of _____ ampere capacity shall be installed to supply all lighting outlets and all convenience outlets not otherwise provided for below. Outlets shall be divided as equally as possible between these circuits. In living room and bedrooms, outlets shall be divided between two or more circuits.

9. APPLIANCE BRANCH CIRCUITS — At least
_____ appliance branch circuits of 20-
            (Number)
ampere capacity, 3 wires, shall be installed to supply convenience outlets exclusively in the kitchen, laundry, and dining spaces.

10. SPECIAL-PURPOSE OUTLETS AND CIRCUITS shall be installed as follows:

Circuits for	No. of Wires and Ampere Rating	Termination

11. SIGNAL AND COMMUNICATION WIRING —Complete signalling system, including bell-ringing transformer, push buttons, and audible signal equipment, shall be installed as shown on plans.

9–4 BRANCH CIRCUITING IN RURAL INSTALLATIONS

An example of branch circuits in a rural wiring system is shown in Figure 9-3. Note that the service entrance and the main entrance panel are located at the residence. A feeder (sometimes called a subfeeder) is connected between the power take-off lugs in the main entrance panel and separate entrance switch on a subpanel. This subpanel is centrally located in the barn or outbuilding area that is to be served. Feeder wires are chosen of sufficient size that the voltage drop on the feeder run will not exceed 2% when full current demand is placed on the subpanel. Each branch circuit is individually fused or provided with a suitable circuit breaker. Branch-circuit wires should be of sufficient size so that the voltage drop on a branch-circuit run will not exceed 3% at full load. This is a greater voltage drop than in residential branch circuits, but it is tolerated as a practical compromise because of the comparatively long runs that are often required for barns and outbuildings.

When there is more than one major building to be served, it is advisable to install a separate feeder to each major building. Thus, a feeder may be run to a 60-amp subpanel in the barn, and another feeder run to a 30-amp subpanel in a shed, and still another feeder run to a 30-amp subpanel in a machine shop. Either overhead or underground wiring may be utilized. Figure 9-4 shows an example of overhead wiring from building to building. Note that on long runs, the wire gage may need to be larger than is required for a 2% voltage drop, in order to ensure that the wires will not break when loaded with ice during blizzards.

In the example of Figure 9-4, the main entrance switch is located inside of the house. The feeder is brought through the wall and up the side of the house to the entrance head. No. 8 entrance cable is typically employed. Conduit, armored, or sheathed cable may be utilized. The insulators are installed at least 15 feet above ground and at least 12 inches apart. Drip loops are provided as in conventional service entrances. At the far end of the run, a flange-type entrance head (Figure 9-5) is shown; if desired, a flange-type entrance head can also be used at the starting end of the run in Figure 9-4.

Underground wiring is exemplified in Figure 9-6. Trench cable (dual-purpose plastic-sheathed cable) is commonly employed. The wires should be enclosed in conduit if there is any possibility of mechanical damage. It is desirable to bury underground runs 2 feet below the frost line. It is important to make moisture-tight seals where the conductors run through building walls. It is good practice to use a bare ground wire with two-wire cable, to provide a continuous ground circuit throughout the house and the outbuilding wiring system. Although this is not mandatory in rural wiring systems, it is an advisable safety precaution. A good ground connection can be made

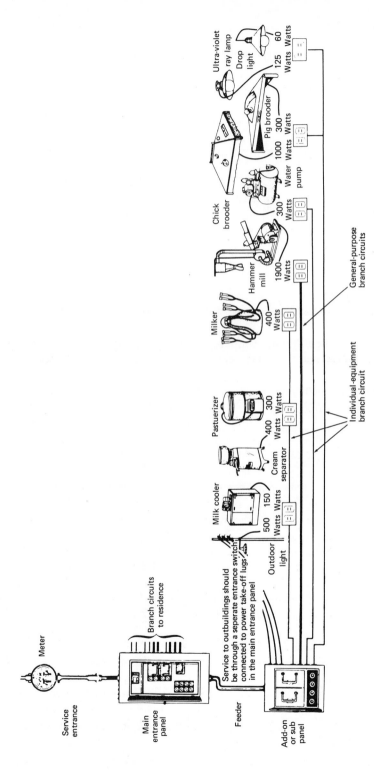

Figure 9-3 Example of branch circuits in a rural installation

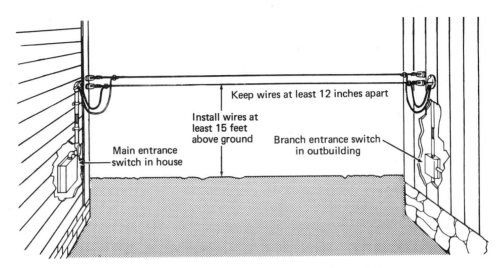

Keep wires at least 12 inches apart

Install wires at
least 15 feet
above ground

Branch entrance switch
in outbuilding

Main entrance
switch in house

Figure 9-4   Example of overhead wiring from building to building

Figure 9-5   Flange-type entrance head

in the house to a cold-water pipe. If a ground connection is made in an outbuilding, it must often be in the form of a ground rod, as noted previously. This is almost always an inferior ground arrangement, compared to a cold-water pipe that has an extensive underground run.

Another example of branch circuits in a rural wiring system is depicted in Figure 9-7. A yardpole (also called a meter pole) is utilized as a central distribution panel for feeder wires to various buildings. As shown in Figure

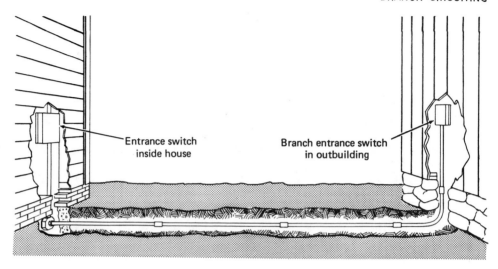

Figure 9-6 An example of underground wiring

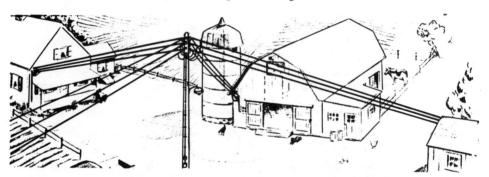

Figure 9-7 Arrangement of a typical yardpole with power wires and feeder wires

9-8, both feeder wires and power wires are connected at the yardpole. The meter is usually mounted on the pole, and connects to the power wires through an entrance head. Feeder wires may connect to the meter through the same conduit as the power wires. However, local codes may require feeder wires to run through a separate length of conduit to the meter. All wires are supported by insulators at least 15 feet above ground, and at least 8 feet from any obstruction, such as a tree or another building.

Water pipes are seldom available in the vicinity of a yardpole, and a ground rod is customarily used, as shown in Figure 9-8. The ground wire should be No. 6 gage connected to the overhead neutral wire at the top of the yardpole. Note that the ground wire must be secured to the pole at 6-inch intervals, or enclosed in conduit or Greenfield if there is any likelihood of mechanical damage. The ground rod should be of copper or other non-rusting substance, at least 1/2 inch in diameter, and 8 feet long. It is driven at least a foot below the soil surface at a point at least 2 feet from the yardpole.

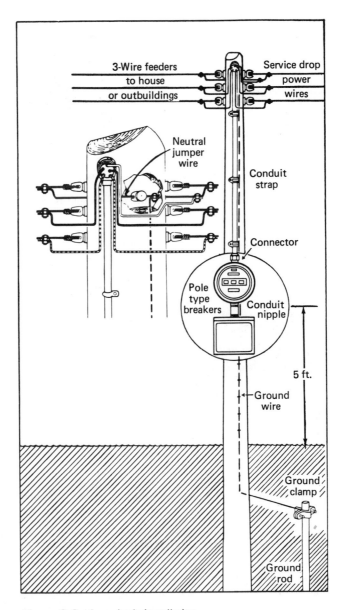

Figure 9-8 A yardpole installation

The entrance switch and breaker are located below the meter on the pole. Note that a split-bolt connector (Figure 9-9) is generally used in connecting the ground wire to the neutral jumper wire.

Figure 9-10 shows how a two-wire service is connected to or tapped off from a three-wire service. This arrangement is used for loads less than 3,500 watts, or motors of 1/2 horsepower or less. Note that a two-wire line may be run from the yardpole, if desired. Each two-wire line has one side grounded

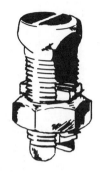

Figure 9-9 A split-bolt connector

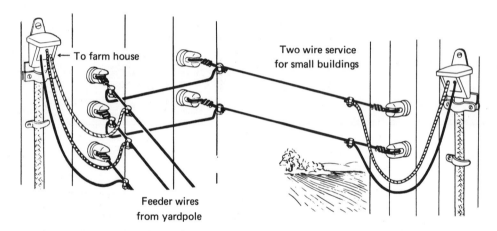

Figure 9-10 Two-wire service tapped into a three-wire service

at the building that is served. No. 8 gage weatherproof wire is generally employed. The wires should be at least 12 inches apart, and installed 10 feet over foot walks, or at least 18 feet over driveways. Either service-entrance cable or conduit may be used to run down the side of the building that is served. A 30-ampere 120-volt service panel is generally installed, with fuses to match the ampacity of the smallest gage wire.

Note that when a good ground is not available, nonmetallic sheathed cable should be used. Also, nonmetallic outlet boxes and switch boxes should be utilized. Porcelain or Bakelite ® sockets are also advised, to minimize the danger of shock. Rural installations are sometimes susceptible to lightning damage. To minimize this hazard, a lightning arrestor should be installed at the yardpole. The arrestor has three leads; the white lead is connected to the neutral wire, and the dark leads are connected to the hot wires. The neutral wire is connected to the ground wire, as explained previously. A lightning arrestor is usually installed at the service switch. On long feeder runs, it is good practice to install another arrestor where the feeder line enters the building that is served.

9–5 VOLTAGE DROPS IN CONDUCTORS

In an ideal wiring system, there would be no voltage drops and the system would operate at 100% efficiency. However, in practice there is inevitably some voltage drop along every wire, and operating efficiency is reduced to some degree. Reduced operating efficiency means that electric energy is wasted as heat in the wires, with a resulting increase in billing charges. Moreover, the voltage drop that occurs in the wires reduces light output from the lamps, reduces power output from motors, and reduces heat output from electric ranges. In turn, system efficiency is reduced as operating costs are increased. As noted previously, every size and type of wire has a certain ampacity rating. This rating represents a practical compromise between efficiency, first cost, and operating costs.

Good practice requires that the voltage drop from beginning to end of any branch circuit is less than 3%. However, a voltage drop not exceeding 2% is preferable. Note that 3% of 120 volts equals 3.6 volts, and that 2% of 120 volts equals 2.4 volts. Again, 3% of 240 volts equals 7.2 volts, and 2% of 240 volts equals 4.8 volts. Although the voltage drop along any length

Table 9-4 Wire gage required for a limit of 2 volts drop in two-wire 120-volt circuits (Courtesy Leviton, Inc.)

Load per Circuit	Current 120-Volt Circuit	Length of run in feet (copper wire)																	
Watts	Amps.	30	40	50	60	70	80	90	100	110	120	130	140	150	160	170	180	190	200
500	4.2	14	14	14	14	14	14	12	12	12	12	12	12	10	10	10	10	10	10
600	5.0	14	14	14	14	14	12	12	12	12	10	10	10	10	10	10	10	8	8
700	5.8	14	14	14	14	12	12	12	10	10	10	10	10	10	8	8	8	8	8
800	6.7	14	14	14	12	12	12	10	10	10	10	10	8	8	8	8	8	8	8
900	7.5	14	14	12	12	12	10	10	10	10	8	8	8	8	8	8	8	8	6
1000	8.3	14	14	12	12	10	10	10	10	10	8	8	8	8	8	8	6	6	6
1200	10.0	14	12	12	10	10	10	10	8	8	8	8	8	6	6	6	6	6	6
1400	11.7	14	12	10	10	10	8	8	8	8	8	6	6	6	6	6	6	6	6
1600	13.3	12	12	10	10	8	8	8	8	6	6	6	6	6	6	6	6	4	4
1800	15.0	12	10	10	10	8	8	8	6	6	6	6	6	6	4	4	4	4	4
2000	16.7	12	10	10	8	8	8	6	6	6	6	6	6	4	4	4	4	4	4
2200	18.3	12	10	10	8	8	8	6	6	6	6	6	4	4	4	4	4	4	2
2400	20.0	10	10	8	8	8	6	6	6	6	6	4	4	4	4	4	4	2	2
2600	21.7	10	10	8	8	6	6	6	6	4	4	4	4	4	4	4	4	2	2
2800	23.3	10	8	8	8	6	6	6	6	4	4	4	4	4	4	4	2	2	2
3000	25.0	10	8	8	6	6	6	6	6	4	4	4	4	4	4	2	2	2	2
3500	29.2	10	8	8	6	6	6	4	4	4	4	2	2	2	2	2	2	2	2
4000	33.3	8	8	6	6	6	4	4	4	4	2	2	2	2	2	2	1	1	1
4500	37.5	8	6	6	6	4	4	4	2	2	2	2	2	2	1	1	1	1	1

and size of wire can be calculated from Ohm's law and knowledge of the resistance of the wire per thousand feet, practical electricians work from standard tables, such as Table 9-4. Note that some local electrical codes forbid installation of wire smaller than No. 12 gage. It is desirable to limit any circuit to a length of less than 100 feet.

9-6 SERVICE CONDUCTOR REQUIREMENTS

Service conductors supply the total current of the branch circuits, and comparatively large conductors must be used to avoid excessive voltage drop. When electric heating is used in a residence, a 200-ampere service is essential. Note that a 100-ampere service is approved only for residences with less than 3,000 square feet of floor area. However, a 60-ampere service may be adequate for a very small dwelling if no major appliances are to be operated. A 30-ampere service can be utilized only for a one-room installation, and no major appliances can be used.

No. 0 wire with RHW insulation is preferred for 150-ampere or 200-ampere service entrances. A three-wire system is used with a neutral conductor so that either 120 or 240 volts are available. No. 2 or No. 3 wire with RHW insulation may be utilized for a 100-ampere service entrance. This is also a three-wire system. No. 6 wire may be installed for a 60-ampere service entrance. A three-wire system is employed. No. 8 wire is generally used for a 30-ampere service. Two conductors are utilized, and only 120 volts are available. These wire gages have been established to reduce voltage drop along the service entrance to a practical minimum under full rated load of the installation.

9-7 ELECTRIC MOTOR CONDUCTORS

Electric motor installation is explained in a following chapter. However, it is instructive at this point to briefly consider the requirements of motor conductors with respect to voltage drops. In general, any electric motor of less than 1/3 horsepower (250 watts) can be operated from a 120-volt outlet; recall that one horsepower is equal to 746 watts. Motors larger than 1/2 horsepower should be operated from a 240-volt outlet, to reduce the voltage drop along the line to the motor. Note that a 1/2 horsepower motor draws approximately 7 amperes from a 120-volt circuit, because a motor is not 100% efficient. A 1/2 horsepower motor draws approximately 3-1/2 amperes from a 240-volt circuit. More importantly, a motor may draw five times its

Table 9-5 More circuit wire gages for individual single-phase motors

Horsepower of Motor	Volts	Approximate Starting Current Amperes	Approximate Full Load Current Amperes	Feet	Length Of Run In Feet from Main Switch to Motor							
					25	50	75	100	150	200	300	400
1/4	120	20	5	Wire Size	14	14	14	12	10	10	8	6
1/3	120	20	5.5	Wire Size	14	14	14	12	10	8	6	6
1/2	120	22	7	Wire Size	14	14	12	12	10	8	6	6
3/4	120	28	9.5	Wire Size	14	12	12	10	8	6	4	4
1/4	240	10	2.5	Wire Size	14	14	14	14	14	14	12	12
1/3	240	10	3	Wire Size	14	14	14	14	14	14	12	10
1/2	240	11	3.5	Wire Size	14	14	14	14	14	12	12	10
3/4	240	14	4.7	Wire Size	14	14	14	14	14	12	10	10
11/2	240	16	5.5	Wire Size	14	14	14	14	14	12	10	10
	240	22	7.6	Wire Size	14	14	14	14	12	10	8	8
2	240	30	10	Wire Size	14	14	14	12	10	10	8	6
3	240	42	14	Wire Size	14	12	12	12	10	8	6	6
5	240	69	23	Wire Size	10	10	10	8	8	6	4	4
71/2	240	100	34	Wire Size	8	8	8	8	6	4	2	2
10	240	130	43	Wire Size	6	6	6	6	4	4	2	1

Table 9-6 Ampacities of various types of extension cords versus length

Ordinary Lamp Cord	Type POSJ SPT	Wire Size No. 16 or 18	Use In residences for lamps or small appliances.
Heavy-duty—with thicker covering	S, SJ or SJT	No. 10, 12 14 or 16	In shops, and outdoors for larger motors, lawn mowers, outdoor lighting, etc.

Types and Applications of Extension Cords

Wire Size	Type	Normal Load	Ampacity
No. 18	S, SJ, SJT or POSJ	5.0 Amp. (600W)	7 Amp. (840W)
No. 16	S, SJ, SJT or POSJ	8.3 Amp. (1000W)	10 Amp. (1200W)
No. 14	S	12.5 Amp. (1500W)	15 Amp. (1800W)
No. 12	S	16.6 Amp. (1900W)	20 Amp. (2400W)

Ampacities of cords (either 2-wire or 3-wire cord)

Light Load (to 7 amps.)	Medium Load (7-10 amps.)	Heavy Load (10-15 amps.)
To 25 Ft. — Use No. 18	To 25 Ft. — Use No. 16	To 25 Ft. — Use No. 14
To 50 Ft. — Use No. 16	To 50 Ft. — Use No. 14	To 50 Ft. — Use No. 12
To 100 Ft. — Use No. 14	To 100 Ft. — Use No. 12	To 100 Ft. — Use No. 10

Gauges Required for Long Cords

rated running current while starting, and the wire size must accommodate this starting current for a short time. To determine wire gages for either portable or permanent motors, the distance from the main switch to the motor and the rating of the motor must be taken into account. Then, the tabulation in Table 9-5 will give the appropriate wire gage.

Lamps and the smaller types of appliances can use No. 16 or No. 18 extension cords. However, if several extension cords are connected in series to operate lamps or an appliance at a considerable distance, it is likely that the lamps will not burn at adequate brightness and that appliances will not operate satisfactorily. This difficulty is due to the excessive voltage drop that occurs along the lengthy extension cord. The only remedy in this situation is to employ extension cords with larger conductors. As would be anticipated, even short cords that supply heavy motors or high-wattage appliances must have comparatively heavy conductors. If an ordinary extension lamp cord were used to connect a heavy motor to an outlet, the cord is likely to burn up. Even if the cord does not ignite, it is unlikely that the motor could be started. Table 9-6 tabulates the ampacities of various types of extension cords, taking the length of the cord into account.

STUDENT EXERCISE

This exercise provides practice in planning branch circuits for residential wiring systems. With reference to Figure 9-11, calculate how many 15-ampere general-purpose circuits will be required, how many 20-ampere kitchen-appliance circuits will be necessary, and how many 240-volt circuits will be required. Assume that 100-watt lamps are to be used throughout.

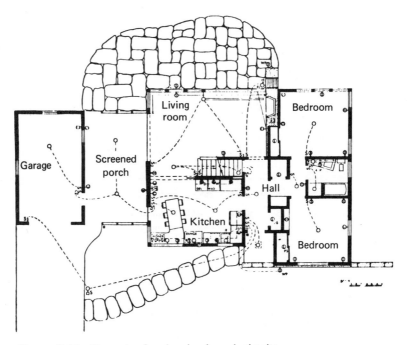

Figure 9-11 Floor plan for planning branch circuits

STUDENT EXERCISE

This exercise provides practical experience with voltage drops along conductors.

1. Obtain a volt-amp-wattmeter, an extension cord, and an electric toaster.

2. Plug the volt-amp-wattmeter into a 117-volt outlet, and plug the toaster into the receptacle provided by the instrument.

3. Measure the line voltage, load current, and real power consumption.

4. Unplug the instrument from the power outlet, and plug an extension cord into the outlet.

5. Plug the instrument into the far end of the extension cord, and plug the toaster into the receptacle provided by the instrument.

6. Measure the line voltage, load current, and real power consumption.

7. Explain why the measured line voltage was greater at the 117-volt outlet than at the end of the extension cord.

8. Explain the difference between the two measured power values, and state where this difference in power was consumed.

9. Did the extension cord become hot?

10. Did the toaster heat up faster or slower when an extension cord was utilized?

11. Suggest how a toaster could be used with a suitable extension cord to avoid excessive voltage drop and loss of efficiency.

STUDENT EXERCISE

This exercise provides practice in evaluating branch-circuit requirements. With reference to Figure 9-16, make these calculations:

1. Five general-purpose 15- or 20-ampere circuits are specified at the top of the diagram. If all of the lamps and appliances shown on these circuits were turned on at the same time, what would be the total power consumption? What would be the total current demand? What is the minimum number of 15-ampere circuits that could accommodate this demand?

2. Two 20-ampere kitchen-appliance circuits are specified in the diagram. If all of the appliances shown on these circuits were turned on at the same time, what would be the total power consumption? What would be the total current demand? Could two 20-ampere circuits accommodate this demand? Is it at all likely that all of the appliances would be operated at the same time?

3. Calculate the maximum power consumption and the maximum current demand for branch circuits 8 and 9. Could the ampacity of either circuit be exceeded if all the indicated loads were turned on at the same time?

4. Calculate the maximum current demand on branch circuits 10 and 12.

5. Calculate the maximum current demand on branch circuit 11.

6. Calculate the maximum current demand for branch circuits 13 and 15.

7. Calculate the current demand for branch circuits 14 and 16.

8. Calculate the approximate current demand for branch circuits 17 and 19.

9. Calculate the current demand for branch circuit 21.

10. Calculate the current demand for branch circuit 22.

11. Calculate the maximum current demand on the service entrance.

12. Would a 200-ampere service entrance be adequate? How do you justify your answer?

STUDENT EXERCISE

This exercise provides familiarity with 3-wire 240-volt receptacles and plugs. Obtain a 50-amp 240-volt (250-volt) receptacle outlet, and a mating 3-prong plug with 14/3 SJ cable.

1. Observe the plug and receptacle, and compare their sizes with that of a 120-volt plug and receptacle.

2. Try the fit of the plug in the receptacle. How many ways can the plug be inserted into the receptacle?

3. Is the color coding of the cable conductors indicated at the terminals on the receptacle?

4. How many conductors are contained in the cable?

5. Is there a grounding wire included in the cable?

6. Note that connections are made to the 240-volt appliance as shown in the diagram below: Black wire to B, red wire to R, and white (or green) wire to G or W.

7. The metal frame of the 240-volt appliance must be grounded to the neutral terminal on the terminal block.

8. Since local codes may require the use of service-entrance cable, for example, instead of 14/3 SJ cable, possible wasted time and expense can be saved by checking with the power company before starting work.

QUESTIONS

1. What are the three basic types of branch circuits utilized in residential wiring systems?

2. Can a heating system operate from a general-purpose branch circuit?

3. In what respects might an incompletely planned wiring system be found unsatisfactory?

4. What is a special circuit?

5. How is a specification form prepared?

6. How do rural wiring systems differ from urban wiring systems?

7. What is meant by the voltage drop along a conductor?

8. How much voltage drop is permissible in a branch-circuit conductor?

9. What is the maximum desirable length for any branch circuit?

10. Why must comparatively large conductors be installed in a service entrance?

11. How much running current does a 1/2 horsepower motor draw? How much starting current?

12. On a long run to a motor, why might unusually large conductors be required, although smaller conductors have adequate ampacity ratings?

13. Why must the length of an extension cord be considered in assigning an ampacity rating?

10 LIGHT AND APPLIANCE SWITCHING CIRCUITS

10–1 GENERAL CONSIDERATIONS

Basic switching circuits were explained previously, so we are now in a good position to consider the switching circuits that are installed by electricians in various situations. With reference to the simple system shown in Figure 10-1, the service-entrance panel provides both automatic and manual switches. If the circuit becomes overloaded for any reason, a circuit breaker will automatically trip and open the circuit. This circuit breaker can also be operated manually to disconnect the branch circuit, and to reconnect it at will. Next, the light switch controls the ceiling light fixture. An automatic time switch controls the post-top lantern. Note that the wall-bracket fixture may have a switch at its base. Or, the fixture may simply be plugged in or out of the receptacle.

To repeat an important principle: all switching is done in the "hot" or black wire. In other words, the neutral or white wire is never open-circuited. Again, to repeat another important principle: black wires connect to black wires, and white wires connect to white wires throughout the system. With reference to Figure 10-1, three white wires and three black wires connect in each junction box—white to white and black to black. Note that the light switch is a single-pole double-throw (SPDT) type. Similarly, the automatic time switch is an SPDT type. A wall-bracket fixture switch is an SPDT type. Note that the circuit breakers in a residential installation are always SPDT types (a solid-neutral installation). On the other hand, the circuit breakers or main disconnect switches in industrial installations may open the "hot" and neutral wires simultaneously, by means of three-pole or four-pole double-throw switches.

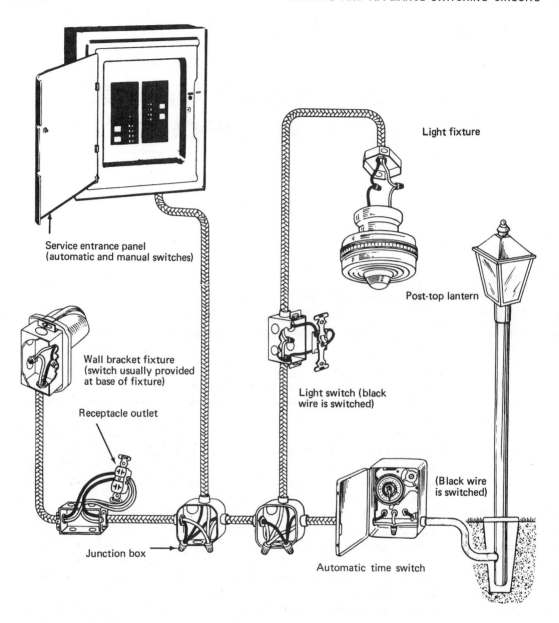

Figure 10-1 Example of manual and automatic switching

10–2 BASIC SWITCHING INSTALLATIONS

If a ceiling fixture is installed with a switch at the end of the run, the circuit is arranged as pictured in Figure 10-2. Observe that the line wires are black and white, and that the white line wire is connected to the white fixture

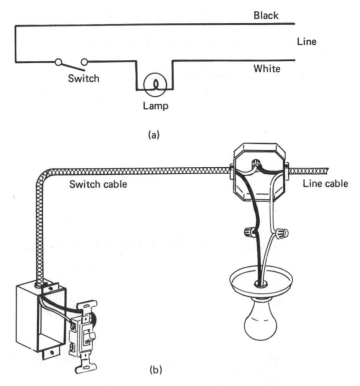

Figure 10-2 Ceiling light controlled by switch at end of run (a) schematic diagram (b) pictorial diagram

wire. Since the cable to the switch has black and white wires, a practical exception is required to the rule that black connects to black and white connects to white. In other words, the switch-cable black wire is connected to the fixture black wire. It is then necessary to connect the switch-cable white wire to the black line wire. In this situation, electricians maintain correct color coding by painting the white switch-cable wire black, both in the switch box and in the light outlet box.

Next, if two ceiling lights are to be installed, one of which is wall-switch controlled and the other pull-chain controlled, the circuit is arranged as shown in Figure 10-3. Note that the black line wire is connected to the black fixture wire, and that the white line wire is connected to the white fixture wire in the pull-chain fixture. In the other fixture, the white line wire is connected to the white fixture wire, but the black line wire does not connect to a fixture lead. A three-wire cable is required from the wall switch to the controlled fixture, and this cable has a black, a white, and a red wire. Observe that the red wire connects to the black wire of the controlled fixture, and to one side of the wall switch; the other side of the wall switch connects to the black line wire.

In case a ceiling outlet is to be controlled by a wall switch, with a conve-

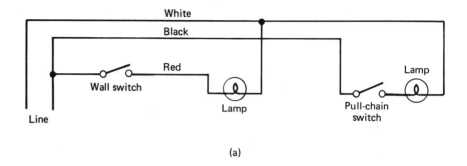

(a)

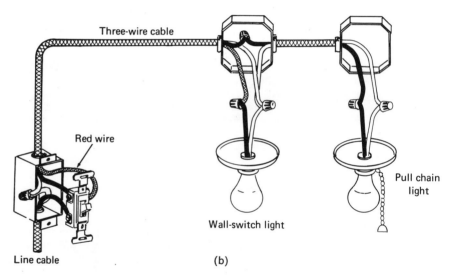

(b)

Figure 10-3 Two ceiling lights, one controlled by a wall switch, the other controlled by a pull chain (a) schematic diagram (b) pictorial diagram

nience outlet at the end of the run, the circuit is arranged as seen in Figure 10-4. Note that the black and white line wires are connected directly to the convenience outlet. A two-wire cable is utilized up to the lamp, and a three-wire cable (black, white, and red wires) is installed between the lamp and the wall switch. From the wall switch, a two-wire cable continues to the convenience outlet. Note that the white fixture wire connects to the white line wire, and that the black fixture wire connects to the red cable wire. This same circuit is employed if the switch and convenience outlet are installed in one outlet box. However, the three-wire cable then terminates at the outlet box, as shown in Figure 10-5.

When a run continues on past a ceiling light and the light is to be wall-switch controlled, the wiring arrangement shown in Figure 10-6 is used. Note

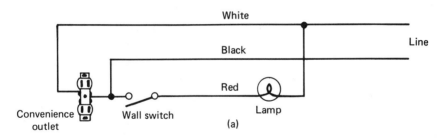

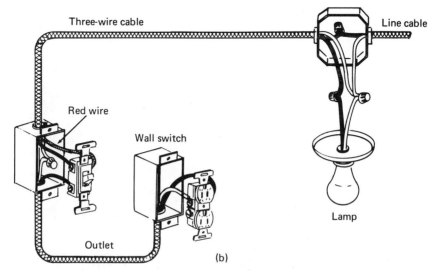

Figure 10-4 Ceiling light controlled by a wall switch with a convenience outlet at the end of the run (a) schematic diagram (b) pictorial diagram

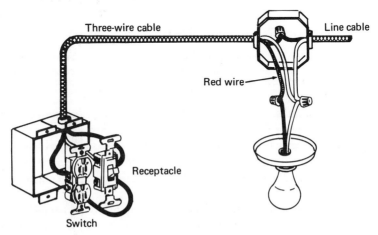

Figure 10-5 Wall-switch controlled ceiling light with switch and outlet in one outlet box

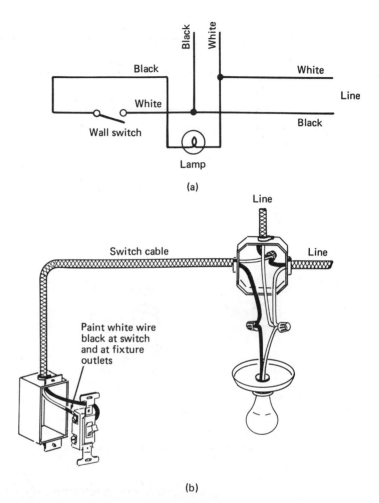

Figure 10-6 Ceiling light controlled by wall switch with line continuing past the ceiling outlet box (a) schematic diagram (b) pictorial diagram

that the line enters the ceiling outlet box and connects to one side of the lamp (white fixture wire to the white line wire). Then the line exits from the ceiling outlet box to continue the run. A two-wire cable is utilized to connect the lamp to the wall switch. Observe that the black cable wire connects to the black fixture wire, and the white cable wire connects to the black line wire. In order to keep the color coding correct, the electrician paints the white cable wire black in both the switch box and the ceiling outlet box.

In case two ceiling lights are to be wall-switch controlled from the same switch box, the wiring connections are made as pictured in Figure 10-7. Note that the white line wire connects to the white fixture wires. A three-

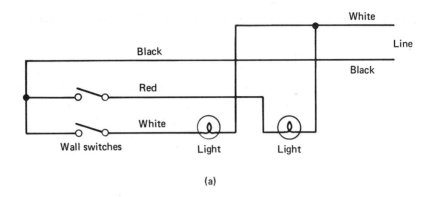

(a)

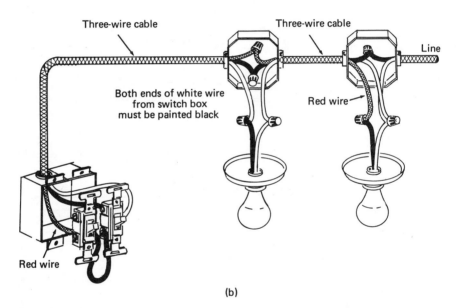

(b)

Figure 10-7 Two ceiling lights wall-switch controlled from the same switch box (a) schematic diagram (b) pictorial diagram

wire cable is installed from the switches to the last ceiling outlet. A three-wire cable is also run between the two ceiling outlets. Observe that the switch-cable black wire connects to both switches and that the red wire connects to one switch, whereas the white wire connects to the other switch. At the last ceiling-outlet box, the white switch-cable wire connects to the black fixture wire. In turn, the black switch-cable wire connects to the red wire of the cable between the outlets. This red wire connects in turn to the black fixture wire in the first ceiling-outlet box. Both ends of the white wire from the switch box to the last ceiling fixture are painted black by the electrician, to maintain correct color coding.

10–3 THREE-WAY AND FOUR-WAY SWITCHING CIRCUITS

As noted previously, three-way switches are often used to control a light from two different locations, so that the light can be turned on or off with either switch. If a light is to be controlled by two three-way switches, both of which are installed past the lamp, the circuit arrangement shown in Figure 10-8 is used. Observe how the black, white, and red wires are connected. A three-wire cable is installed between the two switches. Note that the switch

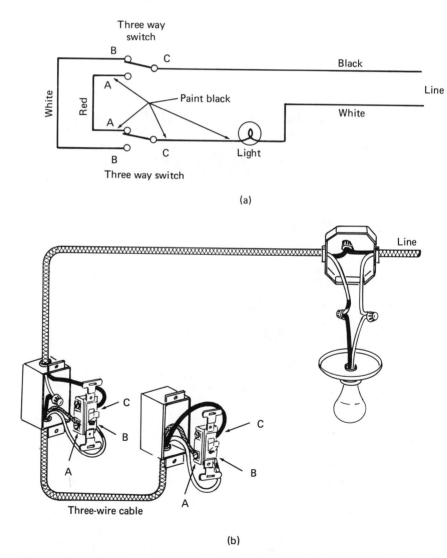

(a)

(b)

Figure 10-8 Light is controlled by two three-way switches, both of which are installed past the lamp (a) schematic diagram (b) pictorial diagram

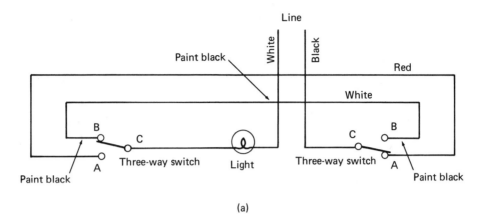

(a)

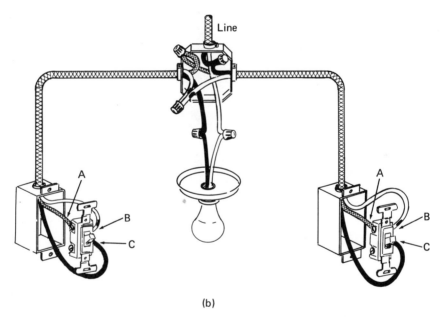

(b)

Figure 10-9 Ceiling outlet installed between a pair of three-way switches (a) schematic diagram (b) pictorial diagram

terminals marked A and B are light-colored terminals. Red and white wires are connected to these terminals. Switch terminal C is a dark-colored terminal that connects to a black wire. To maintain correct color coding, the white wire from the switches is painted black, both in the switch boxes and in the ceiling-fixture box.

In case a ceiling outlet is installed between a pair of three-way switches, the circuit arrangement shown in Figure 10-9 is installed. Note that three-wire cable is utilized between the fixture and each of the three-way switches,

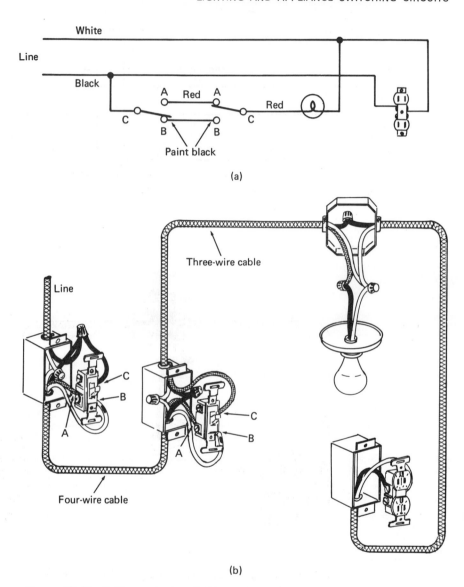

Figure 10-10 Ceiling outlet installed beyond a pair of three-way switches with a convenience outlet (a) schematic diagram (b) pictorial diagram

and that the white wire connects only to the white wire in the fixture. In turn, the black wire connects only to terminal C of the right-hand three-way switch. Also, the red wire connects terminal A of one switch to terminal A of the other switch. As before, the white wire that connects terminal B of one switch to terminal B of the other switch is painted black in each of the boxes. Terminal C of the left-hand switch connects to the black fixture wire.

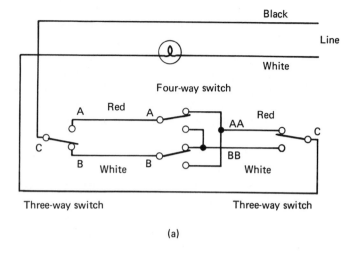

Four-way switch

A — Red — A

C

B — White — B

AA

BB

Red — C

White

Three-way switch   Three-way switch

(a)

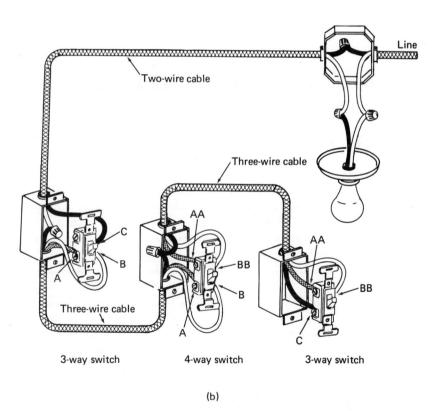

Line

Two-wire cable

Three-wire cable

AA

C

B

BB

A

B

Three-wire cable

AA

A

BB

C

3-way switch   4-way switch   3-way switch

(b)

Figure 10-11 Ceiling outlet controlled from three locations (a) schematic diagram (b) pictorial diagram

If a pair of three-way switches are to control a ceiling outlet which is installed beyond both switches, the circuit arrangement shown in Figure 10-10 is used. In this example, a convenience outlet is located beyond the ceiling outlet. Note that the convenience outlet is not switched and is always "hot." A four-wire cable runs between the two three-way switches, and a three-wire cable runs between the right-hand switch and the ceiling outlet. A two-wire cable runs from the ceiling outlet to the convenience receptacle. It is permissible to install a pair of two-wire cables between the switches, instead of a four-wire cable, if desired.

When a ceiling outlet is to be controlled from three locations, the circuit arrangement pictured in Figure 10-11 is employed. Two three-way switches and one four-way switch are installed. Two-wire cable is run from the ceiling outlet to the first three-way switch. In turn, three-wire cable is run from the first three-way switch to the four-way switch, and from the four-way switch to the second three-way switch. Additional four-way switches can be connected in series with the first four-way switch, if desired, to provide additional control locations. Observe that the white wires from the switches must be painted black in each switch box, and also in the ceiling-outlet box, to keep the color coding correct.

10–4 SWITCH-CONTROLLED SPLIT RECEPTACLES

As noted previously, split receptacles may be switched or may be "hot" all the time. When necessity arises, "hot" receptacles can be switch-controlled in the sense that the circuits can be opened by means of the entrance switch, as pictured in Figure 10-12. In this type of split-wired receptacles, there are two circuits in each box. In other words, a common neutral is utilized with the red wire connected to the top receptacles and the black wire connected to the lower receptacles. By splitting the loads in this manner, the circuit breakers are less likely to trip when several appliances are plugged into the receptacles. No. 12 wire is generally used with the red wire connected to the brass-colored terminals of the top receptacles, the white wire connected to the silver-colored terminals of the upper and lower receptacles, and the black wire connected to the brass-colored terminals of the lower receptacles.

Next, Figure 10-13 shows a split-receptacle circuit in which the upper receptacles are toggle-switch controlled, while the lower receptacles are "hot" all the time. Both the upper and lower receptacles are energized from the same branch circuit in this arrangement. Note that the white neutral wire is connected to the silver-colored terminals of the lower and upper receptacles, the black wire is connected to the brass-colored terminals of the lower receptacles, and the red wire is connected to the brass-colored terminals of

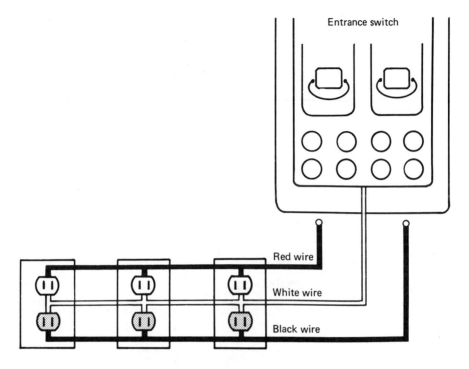

Figure 10-12 Two-circuit duplex receptacles controlled by entrance switch only

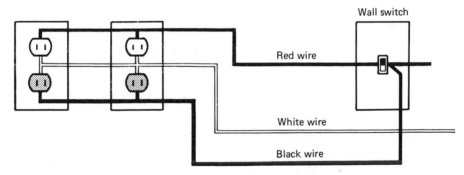

Figure 10-13 Upper receptacles are toggle-switch controlled, and lower receptacles are "hot" all the time

the upper receptacles. This arrangement is often installed when floor lamps are to be controlled by a wall switch.

If comparatively heavy loads are to be provided for, the foregoing arrangement can be energized from two branch circuits, as shown in Figure 10-14. Installation time can be saved if convenience receptacles with break-off tie-

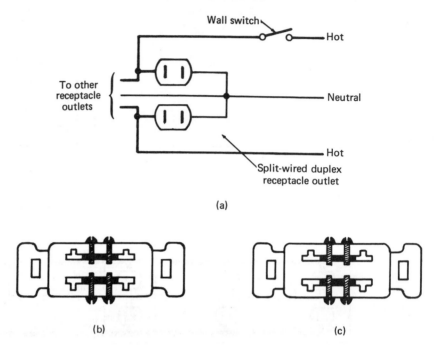

Figure 10-14 Split-receptacle arrangement with two branch circuits (a) circuit diagram (b) single-circuit arrangement of receptacle (c) two-circuit arrangement of receptacle

bar fins between terminals are employed. If the receptacle is to be installed in a conventional single circuit, the fins are not broken off. On the other hand, if the receptacle is to be installed in a two-circuit arrangement, the terminals can be quickly and easily separated by breaking the fins. In an equivalent design, a tie bar is provided between terminals that can be pulled out with a pair of pliers.

10–5 SWITCH AND PILOT-LIGHT CIRCUITING

When a pilot light is to be installed with a switch, the circuit arrangement shown in Figure 10-15 is utilized. Because the pilot light is connected in parallel with the controlled light, both lights are switched on or off simultaneously. This arrangement requires a run of three-wire cable between the pilot light and switch, and to the controlled light. Note that a standard size box can be used for any combination of two or three devices by utilizing interchangeable devices, as shown in Figure 10-16. Thus, a toggle switch, pilot light, and convenience receptacle may be mounted easily and quickly in the same outlet box. Or, a toggle switch and a pair of split receptacles may be mounted in the box.

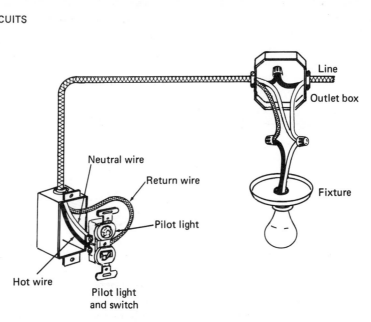

Figure 10-15 Ceiling outlet controlled by a wall switch with a pliot light

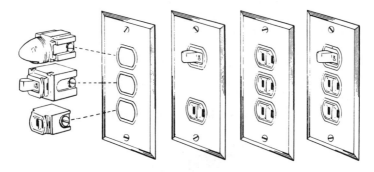

Figure 10-16 Interchangeable devices save installation time with improved appearance

10–6 HEAVY-APPLIANCE CIRCUITS

Branch circuits, outlets, and switches for kitchen appliances were discussed previously. We are now in a good position to consider the more demanding requirements of major appliances. Since heavy appliances have a comparatively large current demand, correspondingly large-gage cable must be installed. Note that a separate three-wire (special-purpose) circuit is required for operating an electric range, an electric water heater, an electric dryer, and the larger types of air conditioners. A range, water heater, and dryer

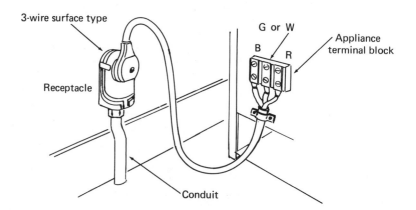

Figure 10-17   Typical range or dryer receptacle installation

can be operated from a 100-ampere service. Local electrical codes often specify the type of cable that must be used in major-appliance circuiting. Service-entrance cable with a bare neutral wire is permitted by the NEC for electric ranges and clothes dryers, but not for other types of heavy appliances.

An electric range is generally installed with No. 6 cable from a separate 50-ampere branch circuit at the main service-entrance panel. A three-wire circuit is required, inasmuch as a range operates at 120 volts for low heat and 240 volts for high heat. Note that the "hot" wires are color-coded black and red, with a white or green neutral wire. Figure 10-17 shows a typical range or dryer receptacle installation, with the "pigtail" cord connected to the appliance terminals. No switch is provided in this example and the appliance is disconnected by unplugging the cord. It is essential to ground the metal frame of the appliance to the neutral terminal, so that a shock will not be

(a)   (b)

Figure 10-18   (a) Typical range receptacles (b) surface-mounted receptacle (Courtesy Leviton Inc.)

received by the user in case the appliance becomes defective. Typical range plugs and receptacles are shown in Figure 10-18. An electric range may draw less than 8,000 watts, but up to 12,000 watts for large units.

Figure 10-19 shows the switching connections for a typical electric-range heater unit. This arrangement provides five steps of temperature. The two heater coils employed may be energized singly or simultaneously. These coils may be connected in series or parallel across the 120- or the 240-volt lines. In the simmer position, the coils are connected in series across 120 volts.

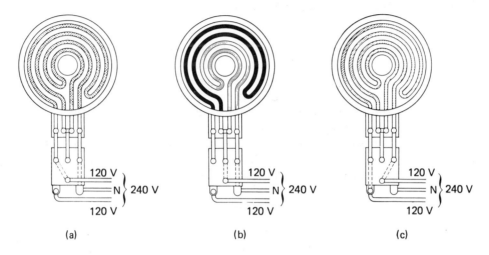

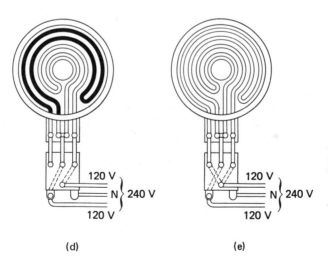

Figure 10-19 Switching connections for a five-step range heater unit (a) simmer circuit (b) low-heat circuit (c) medium-low circuit (d) medium-high circuit (e) high-heat circuit

Next, in the low position, one coil is connected across 120 volts and the other coil is switched out of the circuit. Then, in the medium-low position, both coils are connected in series across 240 volts. In the medium-high position, one coil is connected across 240 volts and the other coil is switched out of the circuit. Finally, in the high position, both coils are connected in parallel across 240 volts.

10–7 ELECTRIC CLOTHES DRYERS

Conventional electric clothes dryers have heating elements that operate on 240 volts, and motors and lights that operate on 120 volts. Accordingly, a heavy-appliance type of receptacle is required, as for an electric range. A separate 30-ampere circuit is generally installed for a dryer that utilizes fuses or switches in the main service-entrance panel. In another arrangement, the dryer circuit is connected at the power-takeoff lugs in the panel, a circuit breaker or fused safety switch is installed near the panel, and conduit is run from this point to the dryer. As in the case of an electric-range installation, the metal frame of the dryer is grounded to the neutral terminal.

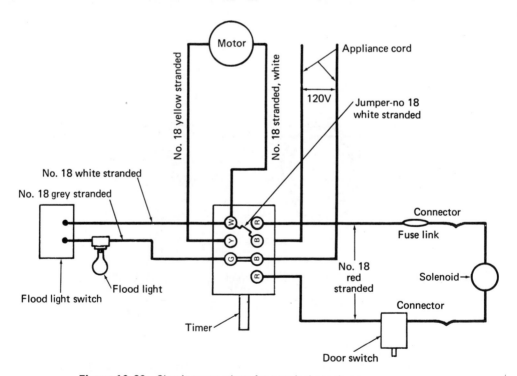

Figure 10-20 Circuit connections for a typical gas dryer

A typical dryer outlet box has a raised ring for mounting in a plaster or sheetrock wall, a flush receptacle, and a plate to cover the ring and box. A conventional dryer draws approximately 4,200 watts, and a high-speed dryer draws about 8,500 watts. In turn, a separate special-purpose 50-ampere branch circuit must be installed for a high-speed dryer. Gas dryers have comparatively simple gas-control circuitry that operates on 120 volts. Figure 10-20 shows the circuit connections for a typical gas dryer. A separate branch circuit is not required for this type of dryer.

10–8 ELECTRIC WATER HEATERS

Water heaters typically draw from 2,500 to 5,000 watts. A single-element heater will have one thermostat and a double-element heater will have two thermostats to control the water temperature. Required methods of wiring are generally specified by local public-utility companies. Figure 10-21 depicts a simple water-heater installation. In this example, the branch circuit is taken from an unfused point in the service-entrance panel and a separate fused safety switch is installed near the panel. A 100-ampere service-entrance panel utilizing circuit breakers is depicted in Figure 10-22. Note that a 30-ampere 240-volt circuit breaker is provided for a water heater, and a 50-ampere 120/240-volt circuit breaker is provided for an electric range.

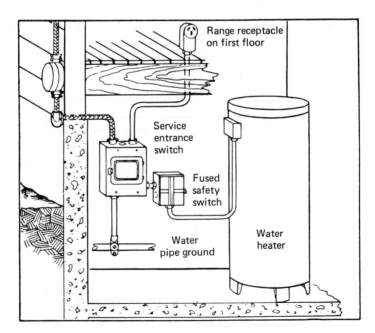

Figure 10-21 Typical water-heater installation

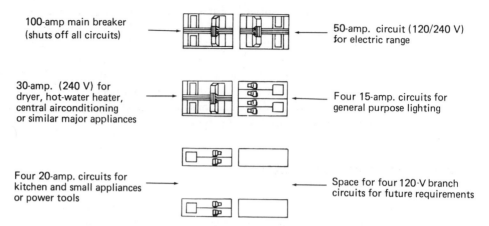

100-amp main breaker (shuts off all circuits) → ← 50-amp. circuit (120/240 V) for electric range

30-amp. (240 V) for dryer, hot-water heater, central airconditioning or similar major appliances → ← Four 15-amp. circuits for general purpose lighting

Four 20-amp. circuits for kitchen and small appliances or power tools → ← Space for four 120-V branch circuits for future requirements

Figure 10-22 Example of a 100-ampere service-entrance panel with circuit breakers for water-heater and electric range branch circuits

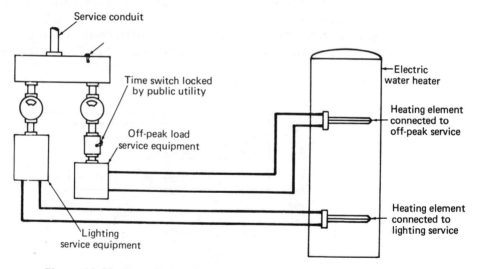

Service conduit

Time switch locked by public utility

Off-peak load service equipment

Lighting service equipment

Electric water heater

Heating element connected to off-peak service

Heating element connected to lighting service

Figure 10-23 Two-element water heater connected to off-peak-load service and to lighting service

As noted previously, power companies often offer a much lower rate for operating water heaters, provided that a separate meter and time switch are installed for off-peak-load operation. Figure 10-23 shows the connections for a water heater with a double-element heater, one of which is energized from the lighting service, whereas the other is energized from an off-peak-load service. This is a compromise installation that takes partial advantage of the lower off-peak-load rate, while ensuring that the hot-water supply will

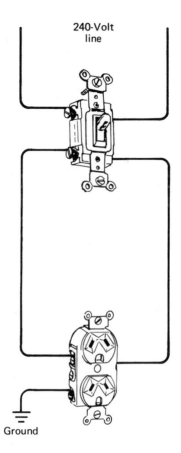

Figure 10-24 DPDT switch is
installed in a 240-volt circuit

never be depleted at any hour of the day or night. Note that double-pole/
double-throw (DPDT) switches are installed in 240-volt circuits, as shown
in Figure 10-24. In turn, the appliance is disconnected from both of the "hot"
wires when the switch is opened. Thus, the neutral wire is not switched and
the appliance remains grounded whether the switch is open or closed.

10–9 AIR CONDITIONERS

Small air conditioners draw approximately 1,000 watts and a 1-ton unit
draws approximately 3,000 watts. If a 1/2-horsepower motor is utilized, for
example, the air conditioner is operated from a 120-volt line. On the other

hand, units with 1-horsepower motors and over are operated from a 240-volt line. Figure 10-25 shows a Westinghouse Unitaire air conditioner that operates in combination with a gas furnace for either cooling or heating a residence, as required by the prevailing temperature.

Both the NEC and applicable local codes must be observed when installing air conditioners. The NEC requires that any exposed noncurrent-carrying parts of an air conditioner that could become "hot" in case of circuit defects must be grounded in an approved manner. As noted previously, a ground connection to a cold-water pipe will always pass electrical inspection. Grounding is required if the exposed parts are within reach of a person standing on the ground outside of the building or if the unit is installed in a hazardous location. Grounding is required even if the air conditioner is connected to metal-clad wiring. Grounding is also required if the unit is installed in a wet location or if it is in contact with metal lath or other metal. In all cases, grounding is required when the circuits operate at more than 150 volts above ground. Note also that the branch-circuit wires must be of sufficient size that the air-conditioner load does not exceed 80% of the conductor ampacity. If lighting units or other appliances are also energized by the branch circuit, the air-conditioner load must not exceed 50% of the conductor ampacity.

Note that three-wire attachment-plug receptacles and caps are available in various forms. Most of them are of the one-position type, in which the plug can be inserted in only one way. These have a grounding-conductor terminal marking and an identified-terminal marking. A grounding-conductor terminal is connected to an independent grounding conductor for the purpose

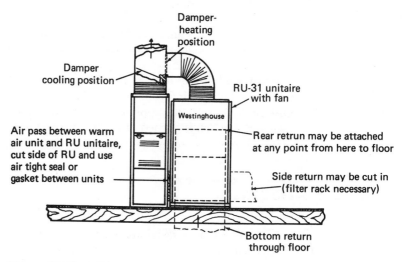

Figure 10-25   A Westinghouse Unitaire® climate-control installation

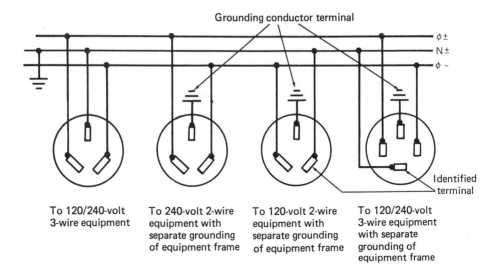

Figure 10-26 Underwriters' Laboratories, Inc. illustration of permissible three-wire receptacle circuiting (Courtesy Underwriters' Laboratories)

of grounding the equipment. It does not normally carry line current to the equipment. On the other hand, the identified terminal is connected to the branch-circuit line which is tapped to the main feeder line having a connection to the grounding terminal. This is the neutral wire, and if the load is unbalanced on either side of the neutral, it will carry more or less line current to the equipment. Figure 10-26 depicts permissible receptacle circuiting in accordance with U-L requirements.

Note that the identified terminal is indicated by means of a metallic-

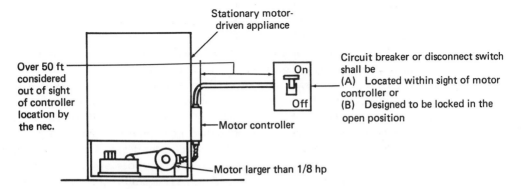

Figure 10-27 Installation of disconnect switch with respect to the motor controller

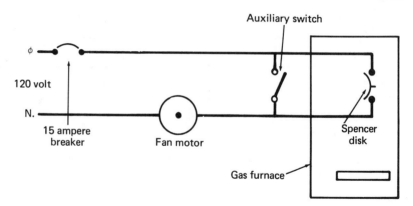

Figure 10-28   Typical circuiting for a motor on a gas furnace

plated coating that is essentially white in color, such as nickel or zinc. Alternatively, the identified terminal may be painted white. On the other hand, a grounding-conductor terminal is marked or stamped with a G, GR, Ground, or Grounding identification on or near the terminal. A receptacle may have an identified terminal but no grounding terminal, or a receptacle may have a grounding terminal but no identified terminal. Again, a receptacle may have both an identified terminal and a grounding-conductor terminal. It is important for the electrician to install the correctly marked type of receptacle in a particular circuit.

Any motor larger than 1/8 horsepower is provided with a controller for starting and stopping the motor, with running-current protection to guard the motor against overload damage and with a disconnect switch to completely disconnect the motor from the line. This controller is mounted on or near the appliance, as shown in Figure 10-27. As noted previously, the running-current protection means is often built into the motor housing. Installation of the disconnect switch or circuit breaker must be made within sight of the motor controller, or the disconnect switch must be of a type that can be locked in its open position. A location over 50 feet from the controller is ruled as out of sight by the NEC.

Figure 10-28 exemplifies a circuit for a fan motor utilized with a gas furnace. Note that the motor is controlled automatically and can also be connected to the line by closing an auxiliary switch manually. A Spencer disk is mounted inside of the furnace and operates as a thermostat in the hot-air stream. All automatic starting devices must be approved for use with the particular motor and must be designed to impose no hazard to personnel when the motor restarts. No running overcurrent device is depicted in Figure 10-28 because the branch circuit breaker provides approved protection. However, a motor rated at greater than 1 HP would require a running overcurrent device.

10–10 BASIC ELECTRIC LAMPS

Nearly all the lamps with which a journeyman electrician is concerned operate from a 120-volt source. Most of these lamps are of the tungsten-filament or fluorescent types. Figure 10-29 shows the bulb outlines and letter designations for some common types of incandescent lamps. Note that S denotes straight, and P denotes pear. Thus, PS indicates pear-straight. T denotes tubular, A denotes arbitrary, G denotes globular, C denotes cone-shape, PAR denotes parabolic, R denotes reflector, and GA denotes globular-arbitrary. Sizes up to 150 watts are commonly used in residential lighting. Night lights in hallways may utilize 7-1/2-watt globular lamps. Driveways are sometimes lighted with 300-watt reflector lamps.

Fluorescent lamps are made varying in length and wattage. The Circline fluorescent lamp is made by General Electric. This circular-type of lamp is used in various fixtures and floor lamps. Note that a fixture is a lighting device that is usually permanently mounted in place, whereas a floor lamp is a portable lighting device. Portable or desk lamps of the high-intensity type use one or more small 6-volt bulbs. A step-down transformer is contained in

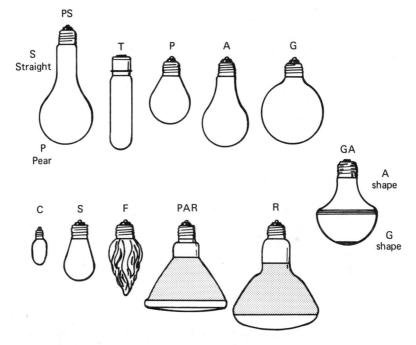

Figure 10-29 Bulb outlines and letter designations for some common incandescent lamps

the base of the lamp to reduce the 120-volt supply to 6 volts. Small neon bulbs are often used for night lights. A neon bulb is operated from a 120-volt line through a current-limiting resistor of approximately 50,000 ohms.

10–11 LAMP SOCKETS

Lamp sockets, also called "lampholders" on official publications, are available in various types. Most tungsten lamps have screw bases of medium size, as depicted in Figure 10-30. However, screw bases of other sizes are also used in fixtures, floor lamps, table lamps, and so on. Most fluorescent lamps use the base arrangements with two pins or a plug. Three different diameters, called miniature, medium, and mogul, are employed. Sockets for screw-base lamps can be grouped into outdoor and indoor types. They can be further grouped into types with and without built-in switches of various designs. Most lamp sockets have terminals for connection to wires. However, some sockets are designed to be plugged into a receptacle.

A basic type of screw-shell socket is commonly called a cleat receptacle. It is designed for indoor application and does not have a built-in switch. Cleat receptacles are available to accommodate the five varieties of lamp bases depicted in Figure 10-30. Cleat receptacles are usually made of porcelain, although plastic is also utilized. Note that the term "cleat receptacle" is not strictly correct, although it is in general usage. In other words, a receptacle is properly a contacting device, as explained above. A plug-type socket may or may not contain a built-in switch. They are approved only for indoor use.

Figure 10-31(a) shows a socket that is related to the cleat receptacle, except that it is designed to be mounted on an outlet box. This is an indoor type of socket and it may or may not contain a built-in switch. Figure 10-31(b) illustrates a brass-shell socket. This type of socket is used indoors; it may be provided with a pull-chain, turn-key, or push-through switch, or it may contain no built-in switch. This type of socket is widely used in lighting fixtures; it is also suspended from a cord over work benches, in sheds, and

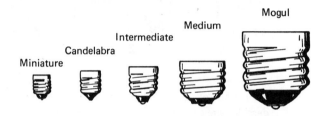

Figure 10-30 Common types of bases for tungsten lamps

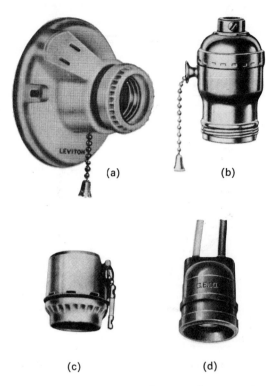

(a)   (b)

(c)   (d)

Figure 10-31 (a) Socket for mounting on outlet box (Courtesy Leviton Inc.) (b) brass-shell socket (Courtesy General Electric Co.) (c) sign-receptacle type of socket (Courtesy General Electric Co.) (d) weatherproof socket (Courtesy General Electric Co.)

similar locations. Another type of socket often used in lighting fixtures is the sign-receptacle type shown in Figure 10-31(c). This type of socket may or may not contain a built-in switch, and is used only indoors. For outdoor installation, a weatherproof socket, such as illustrated in Figure 10-31(d), is utilized.

STUDENT EXERCISE

This exercise provides practical experience with three-way circuiting. A mock-up or "breadboard" arrangement is utilized.

1. Obtain two switch boxes, two three-way switches, a fixture outlet box, several feet of three-conductor nonmetallic sheathed cable, a length of two-conductor nonmetallic sheathed cable, a small lighting fixture, five wire nuts, and a receptacle plug.

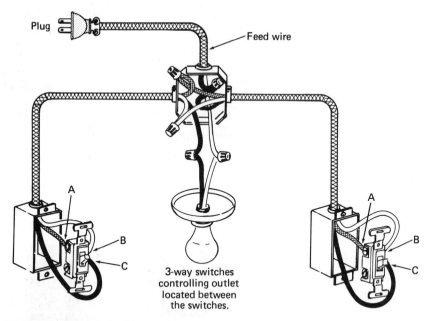

Figure 10-32   Switch wiring project

2.   Arrange the hardware items on a table top as depicted in Figure 10-32.

3.   Cut the cables to suitable lengths and strip the insulation from the ends of the cables.

4.   Make the connections as shown in Figure 10-32.

5.   Connect the black, white, and red wires as shown. Terminals A and B are the light-colored terminals to which red and white wires must be connected. Terminal C is dark-colored, to which the black wire must be connected.

6.   Paint the ends of the white wires from the switches black, both at the switches and at the light outlet.

7.   Plug the feed wire into a 120-volt outlet and check the operation of the three-way switches. If the circuit does not operate correctly, check your connections carefully to locate the wiring error.

STUDENT EXERCISE

This exercise provides familiarity with troubleshooting the wiring defect called "losing the neutral." With reference to the diagram in Figure 10-33, consider the circuit change that occurs when the neutral wire is broken at point X, and answer these questions:

1.   Do lamps L_1, L_2, L_7, and L_8 operate normally when their switches are closed?

2.   If S_3 only is closed, does L_3 light up?

3.   If both S_3 and S_9 are closed, do L_3 and L_9 light up?

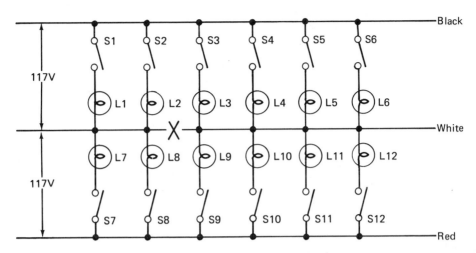

Figure 10-33 Neutral troubleshooting project

4. In case S_3, S_9, S_{10}, and S_{11} are closed, do L_3, L_9, L_{10}, and L_{11} light up?

5. When L_3, L_9, L_{10}, and L_{11} are turned on, do they all glow with the same brightness?

6. Discuss the relative brightness of the lamps when all the switches are turned on, if the upper row of lamps are 100-watt types and the lower row of lamps are 40-watt types.

7. Explain how the location of a break in a neutral wire could be located approximately by means of switching tests.

STUDENT EXERCISE

This exercise provides practice in calculating required feeder ampacity. Make the indicated calculations, and answer the final question.

1. A single-family residence has a floor area of 1,500 square feet, exclusive of an unoccupied cellar, an unfinished attic, and open porches. A 12-kW range is installed.

2. The general lighting load is calculated at 3 watts per square foot:

$$1,500 \times 3 = \underline{\hspace{2em}} \text{ watts}$$

3. The minimum number of branch circuits required, according to NEC regulations, is:
   General lighting load, $4,500/115 = 39.1$ amps; three 15-amp two-wire circuits will be required, or two 20-amp two-wire circuits.
   Small appliance load will require two 20-amp two-wire circuits.
   Laundry load will require one two-wire 20-amp circuit.

4.   The required feeder ampacity is determined as follows:
        Calculated load

	Watts
General lighting	4,500
Small appliance load	3,000
Laundry	1,500
Total (without range)	9,000

Of this calculated load, the NEC stipulates that the first 3,000 watts shall be calculated at 100%, and the remainder at 35%.
Or, the net calculated power demand (without the range load) is:

$$3,000 + 2,100 = \underline{\qquad} \text{ watts}$$

Next, the NEC stipulates that the 12-kW range load shall be calculated at 8,000 watts.
In turn, the net calculated feeder power demand becomes:

$$5,100 + 8,000 = \underline{\qquad} \text{ watts}$$

This power demand corresponds to an ampacity of:

$$13,100/230 = \underline{\qquad} \text{ amperes}$$

Will a 100-ampere service be adequate, or must 200-ampere service be installed?

QUESTIONS

1.   Is switching done in the "hot" line or the neutral line?
2.   What is a solid-neutral installation?
3.   When are the ends of white wires painted black?
4.   How many three-way switches may be used in any one circuit? How many four-way switches?
5.   What kind of a circuit would require three-wire cable? Four-wire cable?
6.   Does a white wire connect to a silver-colored terminal or to a brass-colored terminal?
7.   What is a break-off tie-bar fin?
8.   How are interchangeable devices defined?
9.   What two types of heavy appliances can be wired with service-entrance cable?
10.   How much power does a typical electric range consume?
11.   How much power does a typical dryer consume?
12.   How much power does a typical water heater consume?
13.   What is meant by off-peak-load service?

14. What is a one-position three-wire attachment plug?

15. How many basic bulb outlines are there for incandescent lamps?

16. What is a cleat receptacle?

17. What is a push-through switch?

11 INCIDENTAL CIRCUITING AND INSTALLATION

11-1 GENERAL CONSIDERATIONS

Practical electricians are often concerned with signal, remote-control, low-energy power, and low-voltage power circuits and installation. Basic doorbell and chime circuits were noted previously, so we are now in a good position to consider incidental circuiting in somewhat greater detail. Nos. 18 and 16 gage conductors may be used in flexible-cord form, in a cable, or installed in a raceway. Suitable conductors are Types RF-2, FF-2, RFH-2, TF, TFF, TFN, TFFN, PF, PGF, PFF, PGFF, PTF, PTFF, SF-2, SFF-2, or MTW. These wire types are described as follows:

RF-2: Solid or seven-strand rubber-covered fixture wire.
FF-2: Flexible stranded rubber-covered fixture wire.
RFH-2: Solid or seven-strand heat-resistant rubber-covered fixture wire.
TF: Solid or stranded thermoplastic-covered fixture wire.
TFF: Flexible stranded thermoplastic-covered fixture wire.
TFN: Solid or stranded heat-resistant thermoplastic-covered fixture wire.
TFFN: Flexible stranded heat-resistant thermoplastic-covered fixture wire.
PF: Solid or seven-strand fluorinated ethylene propylene fixture wire.
PGF: Same as PF, except that it can withstand higher temperatures.
PFF: Fluorinated ethylene propylene fixture wire.
PGFF: Same as PFF, except that it can withstand higher temperatures.
PTF: Solid or seven-strand extruded polytetrafluoroethylene fixture wire.
PTFF: Same as PTF, but with flexible stranding.
SF-2: Solid or seven-strand silicone rubber-insulated fixture wire.
SFF-2: Flexible stranded silicone rubber-insulated fixture wire.
MTW: Moisture-, heat-, and oil-resistant thermoplastic machine-tool wire.

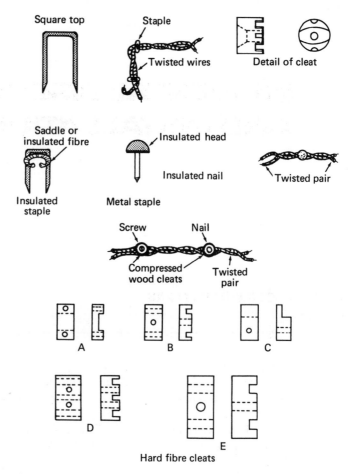

Figure 11-1 Wires may be secured by insulated staples, nails, or cleats

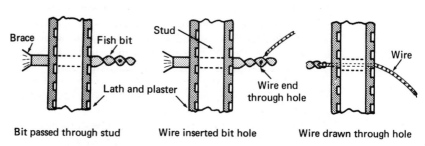

Figure 11-2 How a feeding bit is used

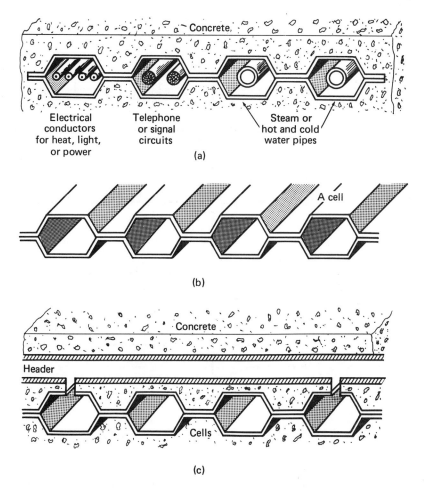

Electrical conductors for heat, light, or power

Telephone or signal circuits

Steam or hot and cold water pipes

(a)

A cell

(b)

Concrete

Header

Cells

(c)

Figure 11-3 Example of a cellular metal floor raceway installation (a) cross-sectional view of raceways (b) designation of a cell (c) a header connecting two cells

Twin-conductor wires or cords are often installed along walls, over or behind baseboards, and over moldings by means of insulated staples or cleats, as depicted in Figure 11-1. The conductors may be fished behind walls in old work. It is occasionally helpful to employ a feeding bit, as shown in Figure 11-2. After a hole is bored through a stud, and before the feeding bit is withdrawn, the wire is inserted in a hole provided in the bit. Then, when the bit is withdrawn, it pulls the wire through the hole. A feeding bit is also called a fish bit.

A raceway is any channel that has been designed for holding wires, cables, or bus bars in accordance with the National Electrical Code. Raceways are

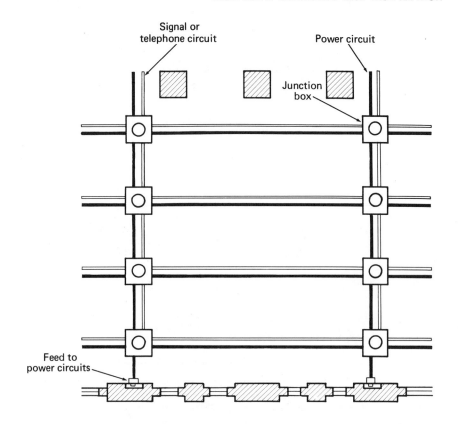

Figure 11-4 An underfloor raceway installation

made of metal or of insulating material. A raceway may consist of rigid metallic conduit, rigid nonmetallic conduit, flexible metal conduit, electrical metallic tubing, underfloor raceways, cellular concrete floor raceways, cellular metal floor raceways, surface raceways, structural raceways, wireways, and busways. A cross-sectional view of typical cellular metal floor raceways is shown in Figure 11-3. An individual raceway is called a cell, and a header is a run of conduit that connects two or more cells. Figure 11-4 depicts an underfloor raceway installation.

More than one pushbutton may be installed for a doorbell, as shown in Figure 11-5(a). If two bells and two pushbuttons are utilized in a three-wire circuit, a return-call arrangement can be employed. Single-pole single-throw (SPST) pushbuttons can be used, as depicted in Figure 11-5(b), or double-pole double-throw (DPDT) pushbuttons can be used, as shown in Figure 11-5(c). When a more subdued signal is desired, buzzers are used instead of bells. Two-tone chimes are also installed with a three-wire circuit, as shown in Figure 11-6. Conventional bell-ringing transformers are utilized

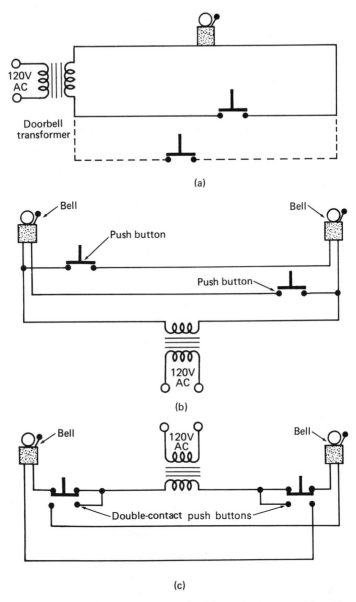

Figure 11-5 Typical doorbell circuits (a) one bell operated by either of two pushbuttons (b) return-call circuit, using SPST pushbuttons (c) return-call circuit, using DPDT pushbuttons

to provide an open-circuit secondary voltage of approximately 16 volts. The current capability of the transformer is such that it will not be damaged if the secondary is short-circuited.

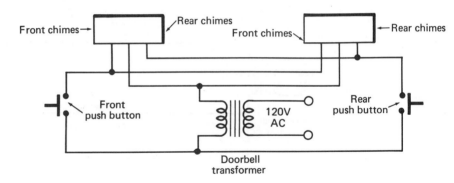

Figure 11-6 Three-wire circuit for two-tone chimes

11–2 REMOTE-CONTROL CIRCUITING

When lights are to be switched from a number of locations, wiring installations may be simplified by using remote control. Relays are utilized to switch the lighting circuits. These relays operate on 24-volt AC lines, and may be installed in each lighting outlet or other locations. Remote control systems require less complex wiring and less expensive wiring than if three-way and four-way switches were utilized. The 24-volt circuits are wired with the same types of wires employed in doorbell and chimes circuits.

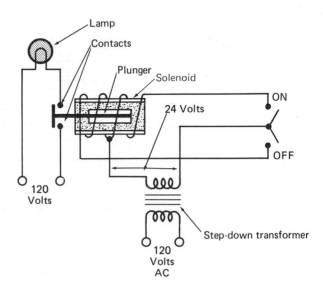

Figure 11-7 Principle of remote-control operation

Figure 11-7 shows the principle of operation. A solenoid is used to move a plunger back and forth, to open or close the lighting circuit. The solenoid is energized by a three-wire 24-volt circuit.

Three basic systems are in general use. The system shown in Figure 11-7 utilizes a central power source with three-wire switching. Note that as many switches as desired may be connected in parallel. Another system uses a central power source with two-wire switching. It employs another type of relay with switches that do not have separate on and off positions. Instead, the switch is pressed once to turn the light on, and is pressed a second time to turn the light off. As in the first system, as many switches as desired may be connected in parallel. In the third system, a local power source is used with three-wire switching. That is, each relay is operated by an individual transformer. The switches have separate on and off positions. As before, as many switches as desired may be paralleled.

Figure 11-8 shows a remote-control circuit in which four switches control two fixtures. Two relays are employed with a single transformer. One of the switches controls fixture No. 2 only. Another switch provides control of both fixture No. 1 and fixture No. 2. The other two switches provide control

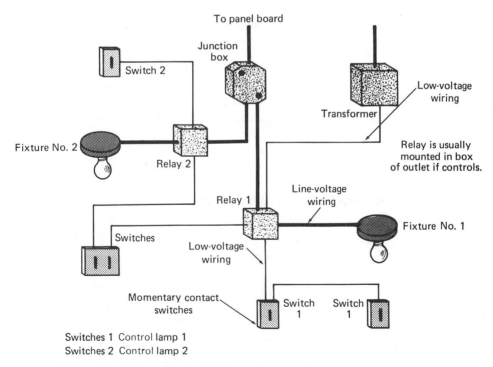

Switches 1 Control lamp 1
Switches 2 Control lamp 2

Figure 11-8 Example of a simple remote-control system

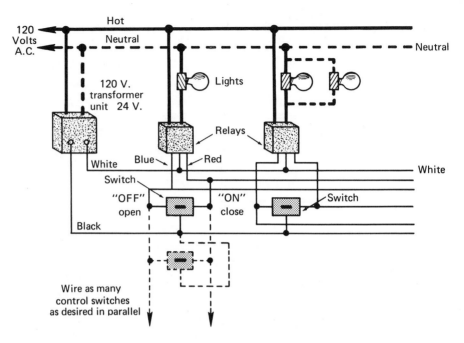

Figure 11-9   Another example of a remote-control system

of fixture No. 1 only. In elaborate installations, a dozen or two dozen switches may be provided in a master bedroom to turn lights on or off both inside and outside the house. Remote-control switches are not usually mounted in boxes. However, when a number of switches are installed at a given location, they may be neatly arrayed in a panel. Labels may be provided for individual switches, to identify the lights that are controlled. Remote-control wires are color-coded, as shown Figure 11-9.

11–3 INTERCOMMUNICATION AND MUSIC UNITS

Intercommunication units installed in residences are usually combined with FM/AM radios that may be supplemented with eight-track tape decks and/or high-fidelity record players. Installation instructions and color-coded wire are generally supplied by the manufacturer. The National Electrical Code stipulates that the connecting wires or cable must not be run closer than 2 inches to any AC line. This applies to AC lines enclosed by conduit, in armored cable, in nonmetallic sheathed cable, and also to low-voltage AC lines. This requirement not only ensures safety in case of electrical system malfunctions, but also minimizes hum pickup by the intercom wires. Similarly, intercom wires must not be run close to and parallel with telephone utility wires, to minimize crosstalk between the systems.

The National Electrical Code also requires that a lightning arrestor be installed on each intercom line whenever there is any possibility of contact with a light or power line due to a support or insulation failure. Note that intercom units are usually of the in-wall type. Other designs are of the on-wall type. Again, the master unit (with the FM/AM radio) may be built into the wall, whereas the substations are on-wall types. Many designs are supported by wallboard or plaster-lath surfaces, although some designs utilize metal boxes. Elaborate master units are comparatively heavy and must be secured to studs or to appropriate wood framing. In addition to substations in various rooms, a front-door substation is sometimes installed. This permits the resident to talk to a caller from any interior substation or the master station.

An intercom and music system is energized by a 120-volt line. Therefore, the electrician must install an outlet where the amplifier is located (usually at the master station). This is typically a permanent behind-the-wall outlet. However, on-wall designs may be powered from a convenience outlet, particularly in old work. Intercom cable runs between master and substations typically employ three conductors. One of these is a common line that is energized with a second line for outgoing audio signals, and also is energized with a third line for incoming audio signals. Concealed wiring is generally utilized, unless brick or equivalent construction in old work prevents fishing wires through walls. In such a case, exposed wiring must be used, and may be run in surface metal raceways both for mechanical protection and for shielding against possible noise and/or hum pickup. In new work, junction boxes are generally installed in the same manner as for light or power outlets. Approximately 3 inches of intercom cable should be left hanging out of the box for final connections after the building is completed.

Typical intercom and audio cables are labelled type A, B, C, and D. Type A has three vinyl insulated conductors. One conductor has a tinned copper shield. The other two conductors are unshielded and the outer insulation is a chrome vinyl jacket. Shielding of the common conductor is occasionally helpful in reducing the noise level without resorting to conduit or metal raceways. Type B has three polyethylene insulated conductors with aluminum Mylar shielding over two of the conductors. The third conductor is unshielded. A No. 20 stranded ground wire is provided and the outer insulation is a chrome vinyl jacket. Type C consists of four polyethylene conductors and is used in the more elaborate intercom/audio installations. One pair of the conductors is tinned-copper shielded. The other pair is unshielded. The outer insulation is a chrome vinyl jacket.

Type D cable is another design of three-conductor cable in which each conductor is aluminum-Mylar shielded. A ground wire is also provided for

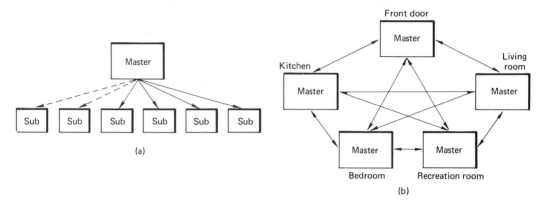

Figure 11-10 Basic types of intercom systems (a) one master station with several substations (b) two-way system throughout, with master stations only

each conductor, consisting of a tinned cadmium-bronze ribbon. This is a comparatively expensive type of cable. Type E is another design of four-conductor cable, employing polypropylene insulation with Beldfoil shielding over one pair of conductors, and is provided with a ground wire. Elaborate intercom/audio installations utilize multiconductor cables, such as type F. This design comprises 12 conductors cabled in three quads (groups of four). Beldfoil shielding is provided over two conductors of each quad, with a clear Mylar shield over the entire quad. The shields are color-coded and provided with a ground wire. The outer insulation is a chrome vinyl jacket.

Two-conductor cable is employed when a one-way intercom system is installed. That is, audio signals originate at the master unit only and are reproduced at any one or all of the sublocations as depicted in Figure 11-10(a). Audio signals cannot be originated at sublocations. On the other hand, three-conductor cable (or multi-conductor cable) is required when a two-way intercom system is installed, as depicted in Figure 11-10(b). In this arrangement, audio signals may originate at any one of the master stations and are reproduced at any one or all of the other master stations. Note that substations can also be installed to supplement a two-way system. Thus, considerable flexibility is available in system planning.

11–4 BURGLAR ALARM WIRING

The National Electrical Code stipulates the same general wiring practices for burglar-alarm systems as for intercom systems. As shown in Figure 11-11, there are two basic types of alarm circuiting. The closed-circuit arrangement employs series-connected switch-contact branches in which the contacts are

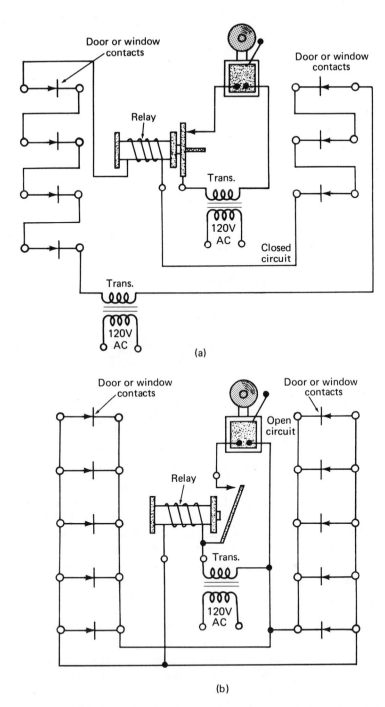

Figure 11-11 Basic burglar alarm systems (a) closed-circuit type (b) open-circuit type

normally closed. When any one contact is opened, the armature of the relay falls back and closes the bell circuit. On the other hand, the open-circuit arrangement utilizes parallel-connected switch contacts that are normally open. If any one contact is closed, the armature of the relay is attracted to the core and closes the bell circuit. Both types of installations are often provided with hand-reset relays, so that the bell will always continue to ring until the armature is unlatched manually.

STUDENT EXERCISE

This exercise demonstrates the characteristics of a doorbell- (or chime-) ringing transformer.

1. Obtain a doorbell transformer, an AC voltmeter, an AC ammeter, and a rheostat of the wire-wound type with approximately 100 ohms maximum resistance.

2. Connect the equipment as shown in Figure 11-12.

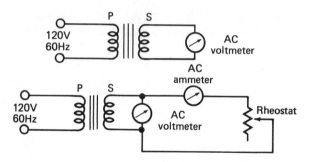

Figure 11-12   Doorbell transformer project

3. With the rheostat disconnected from the secondary terminals, measure the open-circuit secondary voltage with the AC voltmeter.

4. Connect the rheostat and AC ammeter into the secondary circuit, and observe the secondary voltage and current for five different settings of the rheostat as follows: Maximum resistance, 3/4 of maximum resistance, 1/2 of maximum resistance, 1/4 of maximum resistance, and at minimum resistance settings.

5. The change of secondary voltage under increasing secondary current demand is called the regulation of the transformer. If the regulation were 100%, there would be no change in secondary voltage at maximum current demand.

6. Is the regulation of the transformer 100%?

7. Let the transformer operate for 20 minutes with short-circuited secondary terminals. Does the transformer heat up excessively?

8. What is the advantage of poor regulation in a doorbell transformer?

STUDENT EXERCISE

This exercise provides practical experience in wiring doorbell, buzzer, and chimes circuits.

1. Obtain a bell-ringing transformer, doorbell, buzzer, two pushbuttons, a bell-buzzer combination unit, a one- and two-note chimes unit, and a small roll of bell wire.

2. Connect the circuit depicted in Figure 11-13(a). Note that the front button operates the bell, and that the back button operates the buzzer. If the circuit does not operate correctly, check your connections carefully to locate the trouble.

3. Connect the circuit depicted in Figure 11-13(b). Note that A is a common terminal for the bell and the buzzer circuit branches. This arrangement normally works as in the preceding experiment. If the circuit does not operate correctly, check your connections carefully to locate the trouble.

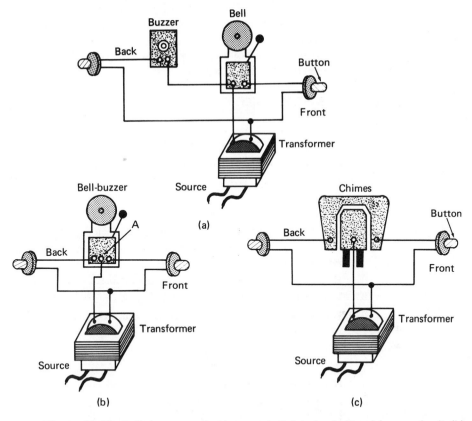

Figure 11-13 Bell, buzzer, and chimes project (a) doorbell and buzzer circuit (b) combination bell-buzzer circuit (c) one-and the two-note chimes circuit

4. Connect the circuit depicted in Figure 11-13(c). Note that the front button operates the two-note chimes, and that the back button operates the one-note chimes. If the chimes are weak, a chimes transformer should be used instead of a bell-ringing transformer. (A chimes transformer provides somewhat higher voltage). If the circuit does not operate correctly, check your connections carefully to locate the trouble.

QUESTIONS

1. What are the ten wire types? Describe each.
2. What is a feeding bit?
3. What is a raceway?
4. What is a cell?
5. What is the name given to a run of conduit that connects two or more cells?
6. What safety factor is designed into a doorbell transformer?
7. What is the approximate open-circuited voltage of most doorbell transformers?
8. What are three methods of switching remote-control circuits?
9. What are the NEC requirements for the spacing between intercommunication cables and power cables?
10. What is the basic difference between an open-circuit and a closed-circuit arrangement for a burglar alarm?

REVIEW QUESTIONS FOR PART II

1. What are some of the symptoms of an inadequate wiring system?
2. How is general illumination usually provided?
3. Why are three-way switches usually required in a stairway lighting circuit?
4. Where would special-purpose outlets be installed in a residential wiring system?
5. Are grounding circuits necessary in farm wiring systems?
6. How is a house-wiring project ordinarily estimated?
7. With reference to Figure II-1, how many switches have been planned? How many fixtures?
8. How many square feet of floor space are allotted to each outlet by the NEC?
9. What is the function of the Underwriters' Laboratories, Inc.?
10. How far above a driveway must service drop wires be installed?
11. What is a hickey?
12. Where are power-takeoff lugs located?
13. Where is a neutral strap located?
14. How is cable "fished" behind walls?
15. Is Greenfield the same as armored cable?
16. What is a surface metal raceway?
17. What is a meter socket?
18. How would you make an ell bend? A saddle bend? An offset bend?
19. How far are receptacle outlets usually mounted above the floor?

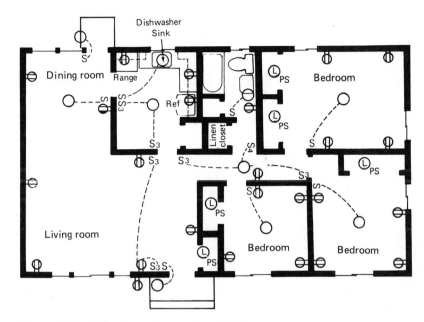

Figure II-1 Wiring layout for a residential living room

20. How far are switches mounted above a floor?

21. How far are wall fixtures mounted above a floor?

22. What is a fixture canopy?

23. At what intervals should conduit be secured to the side of a stud or joist?

24. How many wires may be run into a standard outlet box?

25. What is the purpose of a three-prong grounding plug?

26. What are the three basic types of branch circuits utilized in residential wiring systems?

27. How is a special circuit defined?

28. What is meant by the voltage drop along a conductor?

29. Could an electric range operate from a general-purpose branch circuit? Why?

30. How is a specification form prepared?

31. What is the maximum desirable length for any branch circuit?

32. Is switching accomplished in the "hot" line or the neutral line?

33. What is meant by a solid-neutral installation?

34. How does a three-way switch differ from a four-way switch?

35. What is a one-position three-wire attachment plug?

36. How many basic bulb outlines are there for incandescent lamps?

37. Are all industrial installations of the solid-neutral type?

38. How many major appliances can be operated from a 100-ampere service?

39. Must a separate special-purpose circuit be installed for a high-speed dryer?

40. What is meant by the term off-peak-load service?

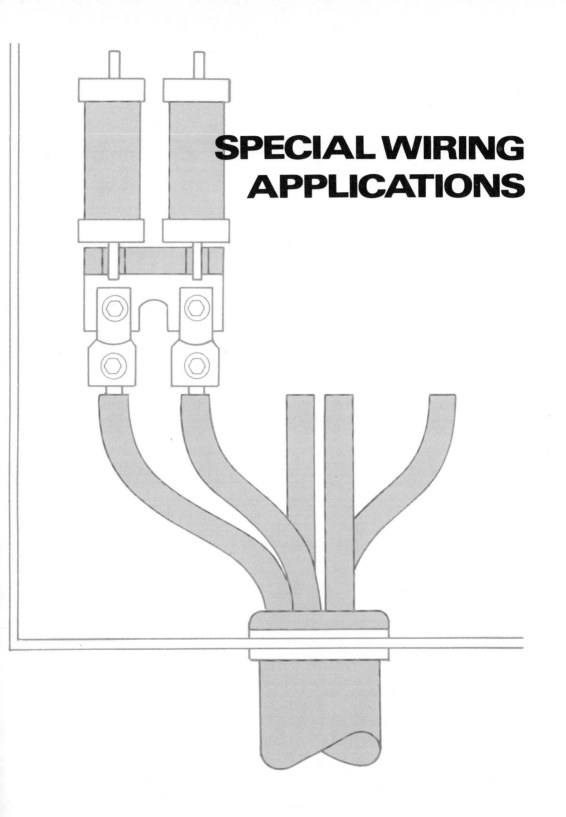

SPECIAL WIRING APPLICATIONS

12 MULTIPLE-OCCUPANCY AND MULTIPLE-DWELLING WIRING SYSTEMS

12–1 GENERAL CONSIDERATIONS

Wiring of an apartment in an apartment house is essentially the same as residential wiring. On the other hand, there are various system requirements to be observed by the electrician in multiple-occupancy and multiple-dwelling wiring systems. Since comparatively heavy loads must often be supplied, larger conductors are generally required for the service entrance than in residential wiring. Required gages of service-entrance wires with respect to various current demands were noted in Chapter 6. Large multiple-occupancy buildings often utilize underground service entrances. Galvanized-iron conduit is usually installed, as noted previously. If conduit is buried in cinder fill, the National Electrical Code requires that the conduit must be encased with at least 2 inches of concrete, as depicted in Figure 12-1(a). However, conduit may be buried directly in the soil, if it is installed at least 18 inches under cinder fill, as shown in Figure 12-1(b).

Only one service is permitted to a multiple-occupancy building, except that a local inspector may give special permission in writing to install more than one service if there is no space available for service equipment that is accessible to all occupants. Note, however, that multiple-occupancy buildings may have two or more separate sets of service-entrance conductors tapped to one service drop or service lateral, as shown in Figure 12-2(a) and (b). The NEC defines a lateral conductor as a wire or cable extending in a general horizontal direction approximately at right angles to the general direction of the line conductors. Again, in the case of larger buildings, parallel service drops may be installed with parallel service-entrance conductors, as in Figure 12-3.

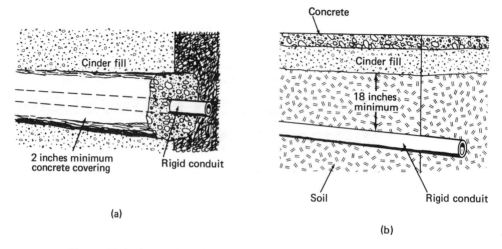

(a)

(b)

Figure 12-1 Installation of conduit underground (a) with concrete covering in cinder fill (b) buried directly in soil below cinder fill

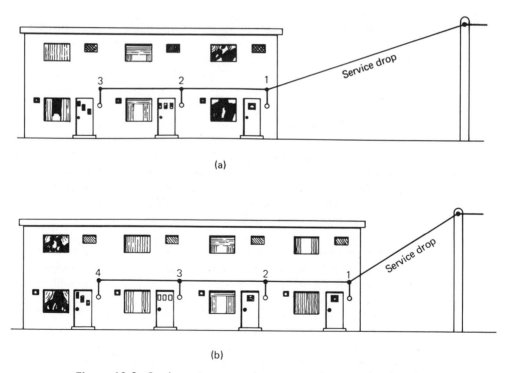

(a)

(b)

Figure 12-2 Service entrance conductors tapped to a service drop (a) three service-entrance conductors fed by a service drop (b) four service-entrance conductors fed by a service drop

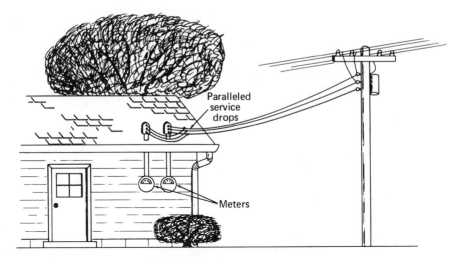

Figure 12-3 Parallel service drops with parallel service-entrance conductors

More than one service may have to be installed if a multiple-occupancy building requires different voltages, different phases, different frequencies, different classes of use such as lighting rate and power rate services, or controlled water-heater service. Service conductors are prohibited from running through one building to another unless the buildings are under the same management. Even in this case the preferred method of installation is with the conductors outside the buildings, as illustrated in Figure 12-4. Note that the NEC stipulates that conductors placed under at least 2 inches of concrete beneath a building are considered to be outside the building. Also, conductors that are installed within a building in conduit or in a duct enclosed by concrete or brick not less than 2 inches thick are considered to be outside the building. A duct is defined as a single enclosed runway for conductors or cables.

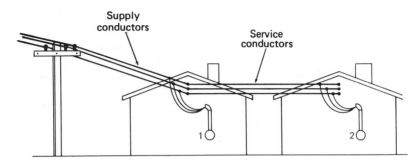

Figure 12-4 Preferred service entrance arrangement for two buildings under the same management

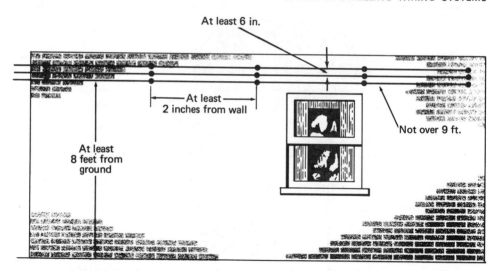

Figure 12-5 NEC requirements for service conductors installed on outside walls

As depicted in Figure 12-5, service conductors installed on an outside wall must be at least 8 feet above ground, supported by insulators not over 9 feet apart, with the wires at least 2 inches from the wall and at least 6 inches from each other. In general, service-entrance conductors must not be spliced, although there are certain exceptions, as follows: clamped or bolted connections in a meter enclosure are permitted by the NEC. Also, as stated previously, buildings of multiple occupancy may have two or more separate sets of service-entrance conductors which are tapped from one service drop or lateral, or two or more subsets of service-entrance conductors

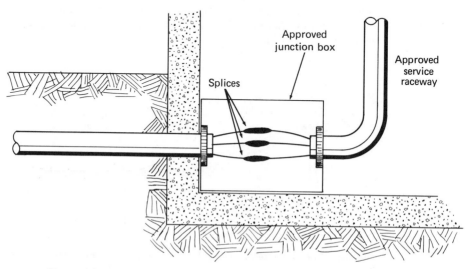

Figure 12-6 Splices are allowed where an underground service conductor enters a building

may be tapped from a single set of main service-entrance conductors. Another exception is made where an underground service conductor enters a building, as in Figure 12-6, where splices may be made in an approved enclosure for connections to an approved service raceway or cable.

12–2 INSTALLATION OF DISCONNECT SWITCHES

In a multiple-occupancy building, each occupant must have access to his disconnect switch or disconnecting means. In addition, a multiple-occupancy building having individual occupancy above the second floor must have the service equipment grouped in a common accessible place, and the disconnecting means must not exceed six switches or six circuit breakers. If a multiple-occupancy building does not have individual occupancy above the second floor, it may have service conductors run to each occupancy, as noted previously. However, each such service must not have more than six switches or circuit breakers. Figure 12-7 shows an installation with six switches. Note that service switch (1) opens the circuits to individual switches (2), (3), and (4), and that service switch (5) opens the circuit to individual switch (6). Note also that a trough is typically an open wood channel with a cover for protecting cables. Figure 12-8 depicts another example.

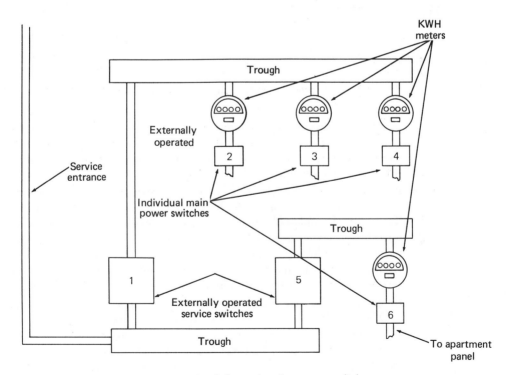

Figure 12-7 An example of six service disconnect switches

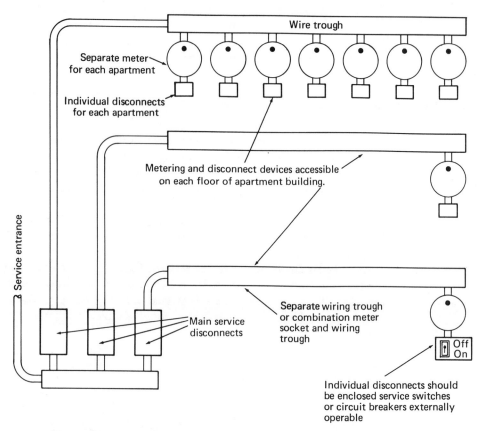

Figure 12-8   Another example of an apartment-house service installation

The disconnecting means must be a manually operable switch or circuit breaker equipped with a handle or equivalent operating provision. In some cases, however, an electrically operated switch or circuit breaker may be utilized, provided that it can be opened by hand in case of power-supply failure. However, an electrically operated switch or circuit breaker need not be capable of being externally operable by hand to the closed position. Two or three single-pole switches or circuit breakers, capable of individual operation, may be installed on multiwire circuits, one pole for each ungrounded conductor, as one multipole disconnect, provided that they are equipped with "handle ties" or a "master handle" to disconnect all conductors of the service with no more than six operations of the hand. Figure 12-9 shows a pair of manually operable circuit breakers with "handle ties."

Of course, it may be necessary to install more than six individual services in a common accessible place. In such a case, a main disconnect switch is installed, as depicted in Figure 12-10. In this example, the main disconnect switch opens the service circuit to a commercial switchboard with eight disconnect switches, eight meter sockets, and eight feeder conduits. These feeder conduits might serve eight apartments, or they could serve seven apartments and a house feeder. In turn, the house disconnect switch might open the circuit to a panel with branch circuits for a washer, dryer, and hall lights. Thus, considerable latitude is permitted by the NEC for planning multiple-occupancy installations. On the other hand, the electrician should never start such an installation without complete plans and without preliminary approval of the local electrical inspector.

Figure 12-9 Example of a pair of manually operable circuit breakers with "handle ties."

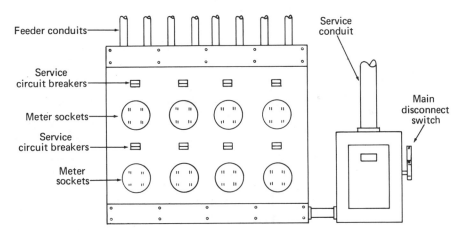

Figure 12-10 Example of a main disconnect switch that controls service to eight feeder conduits

12–3 BRANCH CIRCUITING AND COLOR CODING

Multiple-occupancy buildings, such as apartment houses and hotels, may be supplied by three-phase service consisting of a four-wire installation with a neutral, as shown in Figure 12-11. The voltage between any two of the phase wires is 208 volts, and the voltage between any phase wire and the neutral wire is 120 volts. A three-wire installation is generally made to apartment panels, consisting of two phase wires and the neutral wire. To maintain a reasonably balanced load on the three-phase service, the electrician takes the same number of panelboard taps from each of the three phase wires. Although there will ordinarily be some unbalanced current flowing through the neutral wire, the operating efficiency is acceptable.

Color coding must be observed, as shown in Figure 12-12. That is, the subfeeders or branch circuits that are tapped to the four-wire three-phase

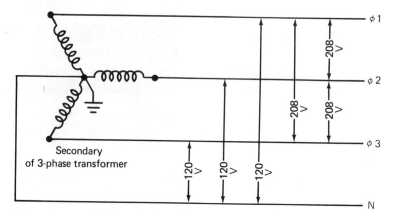

Figure 12-11 A four-wire three-phase service system for a large multiple-occupancy building

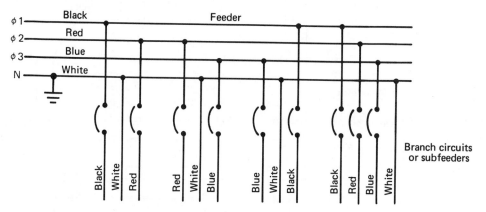

Figure 12-12 Color coding of subfeeder wires tapped to a four-wire three-phase feeder

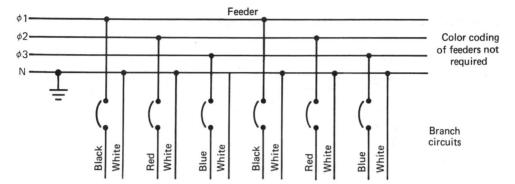

Figure 12-13   Color coding of two-wire branch circuits connected to a multiwire feeder

feeder must be color-coded in accordance with that of the feeder. Each of the three-wire subfeeders will have a white wire, since this is the color of the feeder neutral wire. Similarly, two-wire branch circuits connected to multiwire feeder lines must have the same color wires connected to corresponding feeder wires, as depicted in Figure 12-13. When more than one multiwire branch circuit is installed in a single raceway with branch circuits consisting of fewer conductors than the feeder line, the color coding shown in Figure 12-14 is used. The purpose of this method is to facilitate identification of branch-circuit wiring so that installation can be made with reasonably balanced loads, and to facilitate future maintenance procedures.

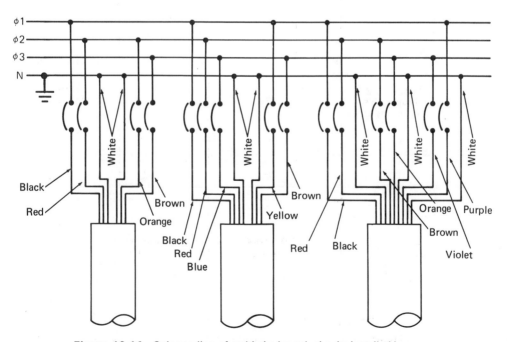

Figure 12-14   Color coding of multiwire branch circuits installed in a raceway

12–4 DUPLEX RESIDENTIAL OCCUPANCIES

The duplex type of residential occupancy shown in Figure 12-15 is quite common in urban areas and in rest homes. Each of the tenants owns or rents one-half of the building and one-half of the yard facilities. In most cases, the tenants are completely separated, as if they occupied completely separate residences. Since one tenant must be disconnected from the service mains when necessary without access to a common accessible location for service equipment, either two service drops must be installed or service-entrance conductors must be extended from one drop to the other half of the building. Inasmuch as such extensions are usually unsightly, and also

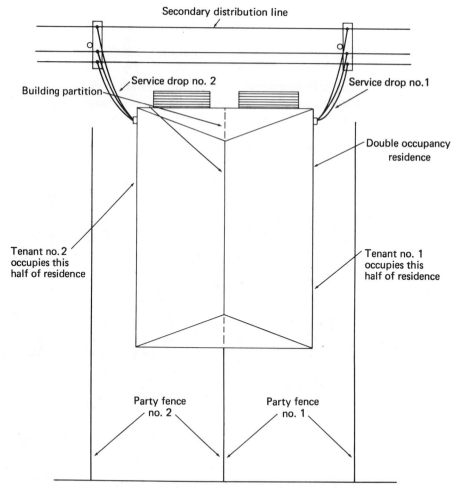

Figure 12-15   Example of two service drops installed in a duplex residential occupancy

comparatively costly, it is preferred to employ two service drops. Note that this requires special permission by the code authority, which is often provided for in municipal ordinances.

12–5 DEMAND FACTOR

In decisions of wire sizes that must be installed, the demand factor of a system may often be taken into account in accordance with provisions of the National Electrical Code. The demand factor is defined as the ratio of the maximum demand of the system, or part of a system, to the total connected load of the system, or of the part of the system under consideration. Since all the lights and appliances in an apartment house will not be turned on at the same time, and since all the heating elements in all the electric ranges will not be operated at the same time, the code permits installation of feeder and branch-circuit wire gages that are somewhat smaller than would be required by the maximum demand of the system.

Note that the demand factors stipulated by the NEC are based on minimum load conditions and 100% power factor. In turn, the electrician may recognize in specific installations that sufficient ampacity is not being provided. Therefore, he should carefully consider the total load that is likely to be connected to the system, so that the installation will pass inspection. The NEC stipulates that the feeder demand factor in dwellings other than hotels shall be taken as 100% for the first 3,000 watts, as 35% for the next 3,001 to 120,000 watts, and as 25% for the remaining load over 120,000 watts.

In hotels and motels, the NEC stipulates that the feeder demand factor shall be taken as 50% for the first 20,000 watts, as 40% for the next 20,001 to 100,000 watts, and as 30% for the remaining load over 100,000 watts. These demand factors also apply to apartment houses without provision for cooking by tenants. However, none of the foregoing demand factors apply to the computed load of subfeeders to areas in hospitals, hotels, or motels where the entire lighting facilities are likely to be used at one time, such as in operating rooms, ballrooms, or dining rooms. In such cases, the total connected load is regarded as equal to the maximum demand. The electrician should carefully evaluate the demand factors of a system in accordance with code requirements before installation is started.

STUDENT EXERCISE

This exercise provides practice for the student in calculating loads and feeder sizes required for multiple dwellings.

Make the indicated calculations and fill in the blank spaces in the following analysis:

Multiple Dwelling

A multiple dwelling has a total floor area of 32,000 sq ft with 40 apartments. The meters are in two banks of 20 each, and individual subfeeders are installed to each apartment. One-half of the apartments are equipped with electric ranges, with a power consumption not in excess of 12 kW each. The area of each apartment is 800 sq ft. The laundry facilities on the premises are available to all tenants. No circuits are to be added to individual apartments. However, we must add 1,500 watts to the house load for each laundry circuit.

Calculated Load for Each Apartment

According to the National Electrical Code, each apartment load is calculated as follows:

General lighting load
800 sq ft at 3 watts per sq ft . . . . .2,400 watts
Special appliance load
Electric range . . . . .8,000 watts

Minimum Number of Branch Circuits Required for Each Apartment

General lighting load: 2,400/115 = ____amps; requiring two 15-amp, 2-wire circuits or two 20-amp, 2-wire circuits.

Small appliance load: Two 2-wire circuits of No. 12 wire as specified by the NEC.

Range circuit: 8,000/230 = ____amps; requiring a circuit of two No. 8 wires and one No. 10 wire.

Minimum Size Subfeeder Required for Each Apartment

Computed load:

General lighting load . . . . .2,400 watts
Small appliance load (two 20-amp circuits) 3,000

Total computed load (without ranges) . . . . .____watts

Application of demand factor:

3,000 watts at 100% . . . . .3,000 watts
2,400 watts at 35% 840

Net calculated load (without ranges) . . . . .____watts

Range load . . . . .8,000 watts

Net calculated load (with ranges) . . . . .____watts

For 115/230-volt, 3-wire system (without ranges):

Net calculated load, 3,840/230 = ____ amps
Size of each subfeeder (see NEC Sec, 215-2)

For 115/230-volt, 3-wire system (with ranges):

Net calculated load, 11,840/230 = ____ amps

Subfeeder neutral:

Lighting and small appliance load . . . . .3,840 watts
Range load, 8,000 watts at 70% . . . . .5,600 watts

Net calculated load (neutral). . . . . .____watts
9,440/230 =____amps

Minimum Size Feeders Required from Service Entrance to Meter Bank
(for 20 apartments—10 with ranges):

Total calculated load:
    Lighting and small appliance load, 20 × 5,400 . . . . .108,000 watts
Application of demand factor:
       3,000 watts at 100% . . . . 3,000 watts
       105,000 watts at 35% . . . . 36,750 _____
               Net calculated lighting and small appliance load . . . . _____watts
    Range load, 10 ranges (less than 12 kW) . . . . 25,000 watts
               Net calculated load (with ranges) . . . . _____watts
For 115/230-volt, 3-wire system:
    Net calculated load, 64,750/230 = _____amps
Feeder neutral:
    Lighting and small appliance load . . . . 39,750 watts
    Range load: 25,000 watts at 70% . . . . 17,500 _____
            Calculated load (neutral) . . . . _____watts
            57,250/230 = _____amps
Further demand factor:
    200 amps at 100% = 200 amps
     49 amps at 70% = 34 _____
          Net load, neutral . . . . _____amps

Minimum Size Main Feeder Required (for 40 apartments—20 with ranges)

Total computed load:
    Lighting and small appliance load, 40 × 5,400 . . . . .216,000 watts
Application of demand factor:
       3,000 watts at 100% . . . . 3,000 watts
       117,000 watts at 35% 40,950
       96,000 watts at 25% . . . . 24,000 _____
            Net computed lighting and small appliance load . . . . . _____watts
    Range load, 20 ranges (less than 12 kW) . . . . 35,000 watts
            Net computed load . . . . _____watts
For 115/230-volt, 3-wire system:
    Net computed load, 102,950/230 = _____amps
Feeder neutral:
    Lighting and small appliance load . . . . 67,950 watts
    Range load, 35,000 watts at 70% . . . . 24,500 watts
         Computed load (neutral) . . . . _____watts
         92,450/230 = _____amps
Further demand factor:
    200 amps at 100% . . . . .200 amps
    202 amps at 70% . . . . .141 _____
         Net computed load (neutral) . . . . ____amps

1.   What are the requirements for installing a service entrance in cinder fill?

2.   What is a duct?

3.   What is the procedure if more than six individual services are installed in a common place?

4.   What is the usual procedure for installing service to a duplex?

5.   What is the demand factor for a system?

6.   What is the demand factor stipulated by the NEC based upon?

13 COMMERCIAL WIRING SYSTEMS

13–1 GENERAL CONSIDERATIONS

Wiring requirements for a small machine shop or cabinet shop are simpler than for an apartment house. On the other hand, a large shop or small factory involves a comparatively elaborate and heavy-duty electrical system. A large factory often installs its own private substation. Special requirements are encountered in wiring chemical plants, in wet locations such as canneries, dairies, and laundries, or in installations where walls are frequently washed down. Severe corrosive conditions must be contended with in pulp and paper mills, tanneries, casing rooms, meat-packing plants, and metal refineries. Explosive vapors occur in oil refineries, and explosive dusts occur in granaries and various mills. Some industrial dusts are of a metallic composition that form conductive films on structural surfaces. All of these conditions affect the wiring requirements of various shops, factories, and industrial installations.

Hazardous locations are grouped into Class 1, Class 2, and Class 3 installations by the NEC. Class 1 locations are those in which flammable gases or vapors are or may be present in the air in quantities sufficient to produce explosive or ignitible mixtures. Class 2 locations are those which are hazardous because of the presence of combustible dust. Class 3 locations are those which are hazardous because of the presence of easily ignitible fibers or flyings, but in which such fibers or flyings are not likely to be in suspension in the air in quantities sufficient to produce ignitible mixtures. Different wiring methods are employed in each class of locations.

13–2 THREE-PHASE SYSTEMS

Three-wire and four-wire Y-connected three-phase circuiting, and other three-phase systems are utilized in industrial installations. Comparatively large currents are generally required, and circuit voltages may also be comparatively high. Three-phase motors are less expensive than equivalent single-phase motors, are more efficient, and are less likely to become defective. Smaller gage wire is also used in three-phase circuits than in equivalent single-phase circuits. Figure 13-1 shows basic single-phase and three-phase lines. Note that there are three "hot" wires in a three-wire three-phase system. There are also three "hot" wires plus a neutral wire in a four-wire three-phase system. The basic distinction between single-phase and three-phase lines is that there is only one alternating current in a single-phase line, whereas there are three alternating currents in a three-phase line.

To understand three-phase motor connections, note that there are 240

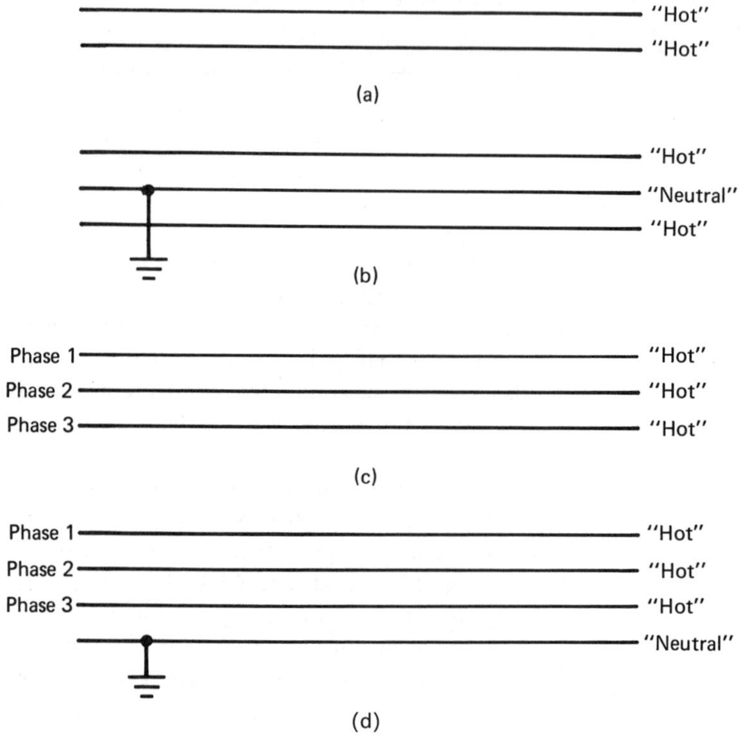

"Hot"
"Hot"

(a)

"Hot"
"Neutral"
"Hot"

(b)

Phase 1 — "Hot"
Phase 2 — "Hot"
Phase 3 — "Hot"

(c)

Phase 1 — "Hot"
Phase 2 — "Hot"
Phase 3 — "Hot"
"Neutral"

(d)

Figure 13-1 Various types of lines (a) two-wire, 120-volt, single-phase (b) three-wire, 120/240-volt, single-phase (c) three-wire, 240-volt, three-phase (d) four-wire, 240-volt, three-phase

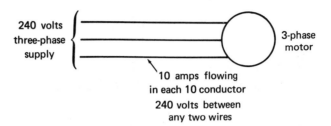

Figure 13-2 Basic three-phase motor connections

volts potential difference between any two wires, as shown in Figure 13-2, and that the same amount of current flows in each wire, such as 10 amperes in the present example. This circuit action results from the timing of the different phases, as seen in Figure 13-3. In other words, the current from the Phase-1 wire goes through its peak or crest value first; next, the current in the Phase-2 wire goes through its peak value; finally, the current in the Phase-3 wire goes through its peak value. While one phase is going through positive values, another phase is going through negative values. Therefore, a potential difference will be measured between any two wires of the three-phase system.

Figure 13-4 shows basic three-phase motor circuits. As pictured in Fig. 13-9(a), the motor has three windings that may be separately energized, although six wires would be required. Note that the Phase-1 winding starts at $S1$ and finishes at $F1$. Next, the Phase-2 windings starts at $S2$ and finishes at $F2$. Then, the Phase-3 winding starts at $S3$ and finishes at $F3$. This is not a practical arrangement because it requires a pair of wires for each phase. Therefore, only three wires are employed in practice for energizing the motor windings. This is done, as shown in (b), by connecting points $F1$, $F2$, and $F3$ together. This connection is called the neutral point. Although a

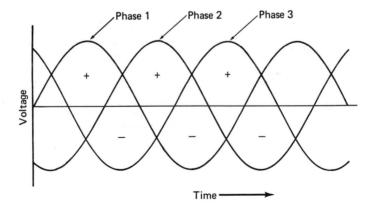

Figure 13-3 Timing of the three voltages in a three-phase system

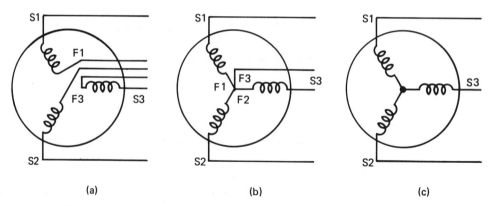

Figure 13-4 Basic three-phase motor circuits (a) three windings, two wires for each phase (b) three windings, one wire for each phase, one neutral wire (c) three windings, one wire for each phase, no neutral wire

neutral wire may be run from this point, no current flows in the neutral wire. Accordingly, the neutral wire is often omitted, as depicted in (c).

When both motors and lights are to be supplied from a three-phase system, it is economical to use a four-wire installation, because the lights can be operated between any one of the phase wires and the neutral wire, as shown in Figure 13-5. This is called a 120/208-volt three-phase system. In other words, there is a 120-volt potential difference between any phase wire and the neutral wire, whereas there is a 208-volt potential difference between any two phase wires. In turn, motors designed to operate in this type of three-phase system are rated at 208 volts. Electricians often call this mode of operation a three-phase network system.

A balanced lighting system results when the lamps are connected as

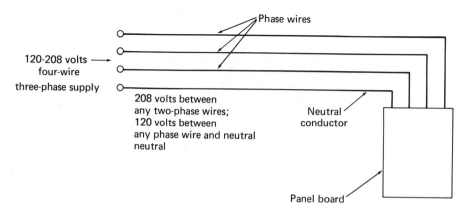

Figure 13-5 A 120/208 volt three-phase installation

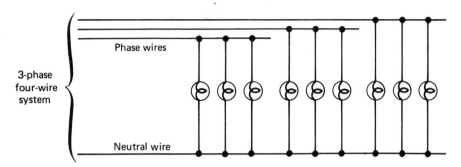

Figure 13-6 Example of balanced light loads in a three-phase four-wire installation

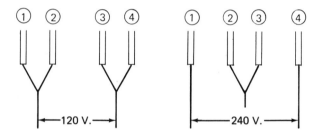

Figure 13-7 Motor terminal connections for 120 and 240 volt operation

shown in Figure 13-6, so that equal currents are drawn from each phase to the neutral wire. In a balanced system, there is no current flowing in the neutral wire. Therefore, there are no I^2R losses in the neutral wire. In practice, however, the system is seldom completely balanced, so that there is more or less current in the neutral wire. In other words, any unbalanced current must flow through the neutral wire. Thus, the chief practical consideration is to connect the lamps so that there will be a reasonable balance in the lighting system. Note that the neutral wire is never larger than a phase wire, and may be smaller in some installations. It is necessary to check any local electrical codes that apply, as well as the National Electrical Code.

Phase rotation, also called phase sequence, is of practical concern to the motor installer. If the phase rotation is in one sequence, a three-phase motor will rotate clockwise. On the other hand, if the phase rotation is reversed, the motor will rotate counterclockwise. Three-phase motor terminals are identified as 1-2-3 or as A-B-C. Similarly, the wires in a three-phase line can be identified. If the 1-2-3 wires are connected to the 1-2-3 motor terminals, the motor will rotate in its standard direction. On the other hand, if wires 2 and 3 are reversed and connected to terminals 3 and 2, the motor will rotate in a reverse direction. A phase-rotation checker is pictured in Figure 13-7(a). When the phase sequence of the wires is 1-2-3 (or A-B-C), as depicted in Figure 13-7(a), the voltmeter reads less than the line-to-line voltage. On the other hand, if the phase sequence is 1-3-2 (or A-C-B), the voltmeter reads more than the line voltage. Note that Figure 13-4 shows the basic Y (wye or star) system.

A phase-sequence tester, such as shown in Figure 13-7(a), employs a resistor, a capacitor, and a voltmeter. Electricians also use a phase-sequence tester arranged as seen in Figure 13-8 with an inductor and two light bulbs. One of the bulbs will glow brighter than the other. If bulb A is brighter than bulb B, the phase sequence is 1-2-3. On the other hand, if bulb B is brighter than bulb A, the phase sequence is reversed, and the motor will rotate oppositely from its standard direction. Still other arrangements of phase-sequence testers are in use, and the electrician follows his personal preference in choosing a tester.

Figure 13-9 shows the basic delta three-phase system. Both the Y and the delta systems are similar in that the three line wires have the same phase relations and the same voltages in corresponding circuits. Note that the delta (Δ) connection is also called a mesh connection. A four-wire Y system provides 120 volts for lighting and 208 volts for power, as noted previously.

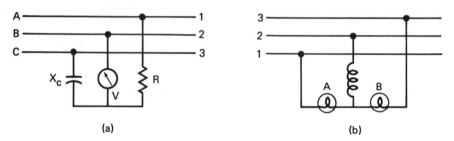

(a)   (b)

Figure 13-8 Phase sequence testers (a) voltmeter reads less than the line-to-line voltage if the phase sequence is 1-2-3 (or A-B-C). (b) bulb A burns brighter than bulb B if the phase sequence is 1-2-3.

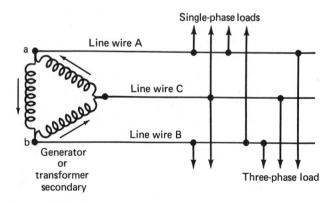

Figure 13-9 The delta three-phase system

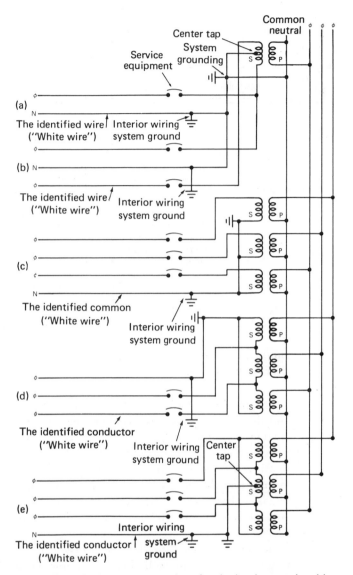

Figure 13-10 Ground connections for single-phase and multi-phase wiring systems (a) two-wire single-phase line (b) three-wire single-phase line (c) four-wire three-phase Y connection (d) three-wire three-phase delta connection with one phase grounded (e) four-wire three-phase delta connections with one phase grounded at a center tap

There is also a four-wire delta system that serves the same purpose, as will be explained.

Various grounding connections are made in industrial wiring systems, as shown in Figure 13-10. A high-voltage three-phase source is stepped down

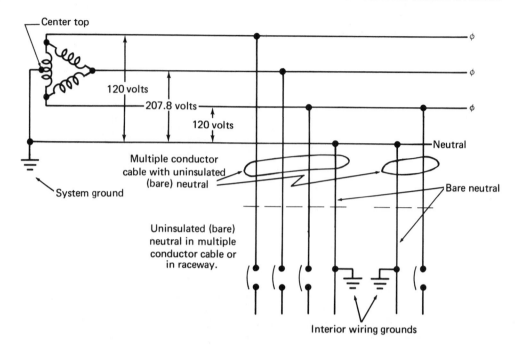

Figure 13-11   When one phase of a delta system is center-tapped, 120-volt and 207.8 volt circuits can be supplied

by means of transformers in this example to supply two single-phase circuits and three three-phase circuits. The single-phase circuits are energized by a single-phase transformer with a center-tapped secondary. Next, the three-phase circuits are energized by three-phase transformers. Observe that the first is a Y-Y connected transformer; the second is a Δ-Y connected transformer, and the third is a Δ-Y connected transformer with one of the secondary phases center-tapped. This center-tap connection provides 120 volts for lighting and 207.8 volts (208 volts) for power, as depicted in Figure 13-11.

In some industrial installations, 277-volt lighting is utilized. This wiring system employs a 480-volt three-phase line which is typically connected in a four-wire Y arrangement. This circuiting provides phase voltages (the voltage between any line wire and neutral) of 277 volts for lighting circuits. An advantage of a 277-volt lighting system is higher efficiency when heavy lighting loads must be installed. Note that the NEC permits industrial branch lighting circuits to be operated as high as 300 volts above ground, provided the lighting fixtures have mogul-base screw-shell lampholders or other approved lampholders. The lighting fixtures must be mounted at least 8 feet above floor level.

13–3 BUSWAYS, WIREWAYS, AND AUXILIARY GUTTERS

Bus duct is used widely to install power circuits that operate at 600 volts or less. It is assembled section-by-section to required lengths for various runs. Bus duct is often less costly than conduit or cable in the long run. It consists of an arrangement of copper bars in a sheet-steel housing. Standardized lengths and fittings are available to meet the requirements of any distribution system. Busways are permitted only for exposed installations and are prohibited where subject to severe physical damage, corrosive vapors, in hoist-

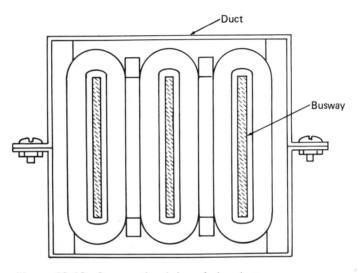

Figure 13-12   Cross sectional view of a bus duct

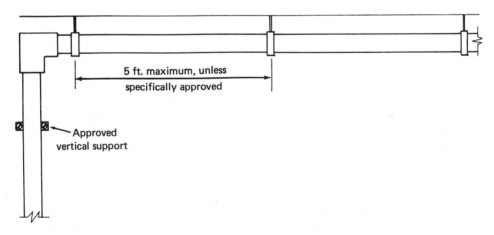

Figure 13-13   Vertical and horizontal support of a busway

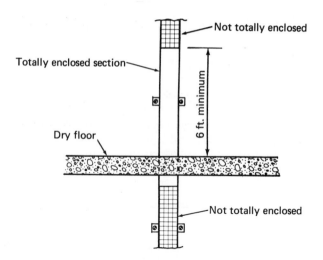

Figure 13-14   Installation of a busway through a floor

Table 13-1   Bus dimensions versus ampere ratings

Ampere Rating	Size of Bus Bar Per Phase Leg (In Inches)	Copper Cross Sectional Area (In Square Inches)	Size of Neutral Bus Bar (In Inches)
225	7/8 x 1/8	0.109	7/8 x 1/8
400	1- 1/2 x 1/4	0.375	1- 1/2 x 1/8
600	2 x 1/4	0.500	2 x 1/8
800	3 x 1/4	0.750	3 x 1/8
1000	4 x 1/4	1.000	4 x 1/8
1250	5 x 1/4	1.250	5 x 1/8
1500	6 x 1/4	1.500	6 x 1/8

ways, or in any hazardous location. Specially approved busways are available for outdoor installation, or in wet or damp locations. Figure 13-12 shows a cross-sectional view of a bus duct.

Adequate support must be provided when installing bus duct. Figure 13-13 depicts the NEC requirements. Horizontal supports are installed at 5-foot intervals unless specific approval has been obtained for 10-foot spacing. Vertical supports must be designed for this particular purpose. Busways may extend transversely through dry walls, provided unbroken lengths of busway are used where passing through the walls. Busways may also extend vertically through dry floors, provided they are totally enclosed (unventilated) where passing through the floors and for a minimum distance of 6 feet above the floor to provide adequate protection from physical damage. This requirement is depicted in Figure 13-14.

Note that there are four basic types of bus duct called plug-in duct, short-run feeder duct, low-impedance duct, and weatherproof duct. Plug-in type is more widely used than the others. It is installed for branch-circuit runs because it is designed with insulated receptacles at 1-foot intervals where power take-offs can be connected. Plug-in duct is also extensively used as feeder duct on short runs or where several tap-offs are made from the feeder between switchgear and other distribution points. Plug-in duct is ordinarily available in current ratings from 225 to 1,500 amperes for single-phase, three-phase, and three-phase four-wire installations. Three-phase four-wire installations usually employ a half-size neutral bus, since the neutral carries only the unbalanced current. However, if a lighting load that is an appreciable fraction of the total power load is also served, a full-sized neutral bus is utilized. Table 13-1 tabulates bus dimensions versus ampere ratings.

Bus bars at one end of the duct are offset, so that when another duct or fitting is mated, the bus bars will join properly. After the two sections have been connected, a cover plate is secured over the opening to enclose the duct. Basic duct fittings are called elbow, tee, cross, offset, unfused reducer adapter, expansion joint, end closer, cable box tap, adapter cubicle, flanged end, and transformer tap. (See Figure 13-15.) Elbows are installed to make right-angle turns. Tees are utilized to make a tap-off from a duct. Crosses are employed to make more than one tap-off at the same location on a duct. Offsets are installed to run duct over or under pipes or building structural members. Unfused reducer adapters are utilized to reduce the ampacity of a duct without employing an overcurrent protective device. Note that the NEC restricts the use of the foregoing fittings to bus bars at least one-third of the rating of the overcurrent protective device back on the line, and the run must not be longer than 50 feet.

Expansion joints are installed to equalize the expansion and contraction of steel and copper during temperature changes. An expansion joint is used in the middle of any run of 100 feet or more. End closers are employed to seal the ends of ducts and to secure bus bars in place. Cable tap boxes are utilized to mate the duct to cables. Figure 13-16 shows how cables are installed with ducts. Adapter cubicles are used where an overcurrent protective device is required for a reduction in bus ampacity of less than one-third of the prior protective device, or if the reduced run is longer than 50 feet. Flanged ends are installed to connect ducts to switchboards, panelboards, or other units. Transformer taps are installed to connect buses to transformer banks.

Short-run feeder duct is similar to plug-in duct, except that it has no provision for installing plug-in devices. It is available in ampacities up to 4,000 amperes and is generally installed for runs of less than 200 feet. Table 13-2 tabulates bus-bar dimensions for short-run feeder bus with 2,000 and

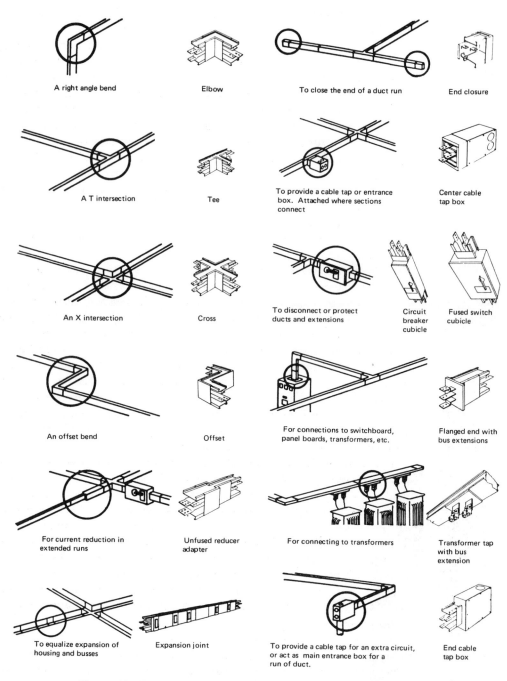

Figure 13-15 Illustration of basic duct fittings (Courtesy Chromalox Inc.)

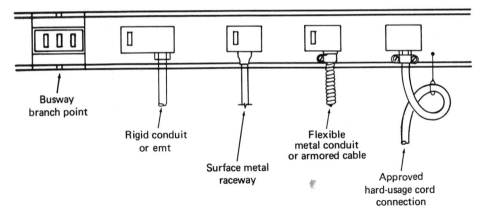

Figure 13-16 Approved methods of installing cables with ducts

Table 13-2 Bus-bar dimensions for short-run feeder duct

Ampacity Rating	Dimensions of Bus Bar Per Phase Leg (In Inches)	Copper Cross Sectional Area (In Square Inches)	Dimensions of Neutral Bus Bar (In Inches)
2000	Two 5 x 1/4	2.5	One 5 x 1/4
2500	Two 6 x 1/4	3.0	One 6 x 1/4

2,500 ampere ampacities. This type of duct often has sufficient inductive reactance that it tends to limit current in the event of a short-circuit fault. Low-impedance duct is installed when heavy currents must be carried on comparatively long runs. It is designed to have a very low value of inductive reactance, to minimize the voltage drop on a long run. Both short-run feeder duct and low-impedance duct are also available in weatherproof design.

When busway systems are operated with grounded conductors, charges of static electricity may build up and flashover will occur, with possible damage to the busway. Therefore, static neutralizers are installed as depicted in Figure 13-17. A static neutralizer consists of high-value resistors, such as 20,000 ohms, that "bleed" static charges to ground. Also, ground detectors are provided, consisting of neon glow lamps. Three lamps are provided to indicate which phase or phases may have developed a ground fault. In normal operation, all three lamps are glowing. However, if a ground fault occurs in one phase, the associated lamp will stop glowing. This device also has a disconnect switch for opening the circuits to the static neutralizer and ground detector.

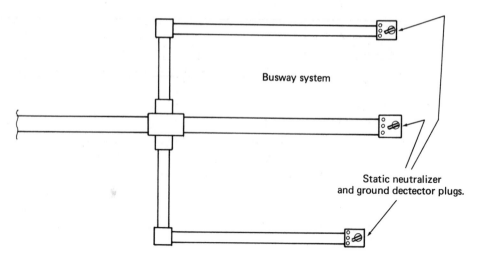

Figure 13-17   Installation of static neutralizers and ground detectors

Wireways have an external appearance similar to busways. However, the conductors are not rigid bus bars, but consist of insulated conductors that pass through the wireway in much the same manner as in conduit. A wireway may be installed with many circuits, whereas a busway is limited to one or two circuits. As many as 30 conductors may be installed in a wireway. Circuit voltages must not exceed 600 volts. A typical wireway is 4″ × 4″ and is formed from 16-gage galvanized steel. Wireways may be installed in dry

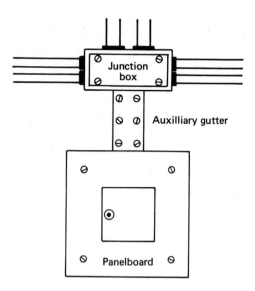

Figure 13-18   Example of an auxiliary gutter

locations and exposed runs, provided they are not subject to mechanical hazards or corrosive fumes. They are generally supported by hangers.

Auxiliary gutters, shown in Figure 13-18, are somewhat similar to wireways except that they do not have hinged covers, but cover plates that are secured by machine screws. Gutters are used principally in short runs of less than 30 feet to supplement wiring spaces at meter centers, distribution centers, switchboards, and similar installations of interior wiring systems. They may enclose either bus bars or insulated conductors, but must not be used to enclose switches, overcurrent devices, or other appliances or apparatus. A gutter must be supported at 5-foot intervals or less. A maximum of 30 conductors may be installed in a gutter, unless conductors are for signaling circuits or motor control circuits.

13–4 MISCELLANEOUS INDUSTRIAL WIRING METHODS

Bars, rods, tubing, or cables may be installed in built-in channels, shafts, or chases of fireproof buildings, as depicted in Figure 13-19. Bare conductors are prohibited in damp, wet, or hazardous locations. If installed in a ventilated channel, 1,200 amperes of current per square inch of bus bar is permitted by the NEC. In unventilated channels, 1,000 amperes per square inch is the maximum current flow that is permitted. These conductors must be supported on noncombustible and nonabsorptive insulating supports of adequate mechanical strength. This type of installation is utilized only for feeders, and special permission must be obtained from the local code authority.

Figure 13-19 Example of bare-conductor feeders installed in a built-in channel

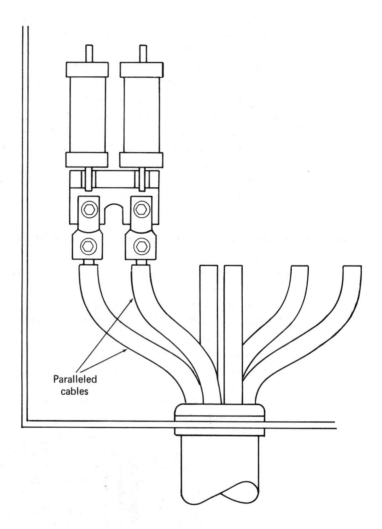

Figure 13-20 Example of parallel (multiple) cable installation

When very heavy currents must be accommodated, conductors may be installed in parallel (multiple), as illustrated in Figure 13-20. Such conductors must be of equal length, have the same gage size, and must have the same type of insulation. Solid bus bars are utilized to secure paralleled cables at their terminal points, to ensure equal division of currents. Aluminum conduit is preferred to iron conduit in heavy-current circuiting because aluminum is nonmagnetic and reduces the losses and the voltage drop of the run.

Note that when wiring is installed for an electrically operated crane, both the frame of the crane and the metal track must be grounded, as de-

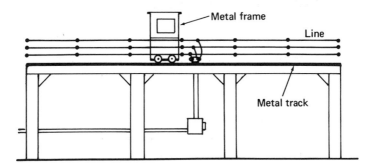

Figure 13-21   Frame and track of crane must be grounded

picted in Figure 13-21. Some industrial elevators are hydraulically driven; if electrical conductors for lights or signals are attached to the elevator car, the metal car frame must be grounded. In the case of an electric elevator, the hand-operated metal shifting rope or cable must be grounded. In electric substations or vaults under sole control of the utility company, metal partitions, grill work, and guard rails are not mandatory. On the other hand, if the substation is privately owned by the industrial concern, these metal structures must be grounded. Also, any indoor or outdoor substation not owned by the utility company must be adequately protected from lightning by installation of lightning arrestors, as shown in Figure 13-22.

All wiring installations and control equipment must be dust-tight in locations where metallic or other electrically conducting dusts are present. If explosive vapors are or may be present in a factory or shop, an explosion-proof panelboard must be installed. Although the selection of such equipment is not ordinarily within the province of the electrician, he should know the code requirements in such matters so that he can catch installation errors before inspection and thereby avoid costly reworking procedures.

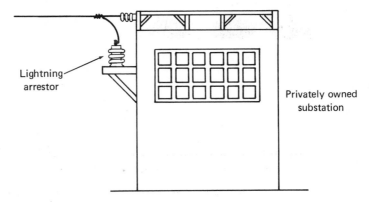

Figure 13-22   Installation of lightning arresters

13–5 LIGHTING SYSTEM PLANNING

As noted previously, an electrician does not always work from plans prepared by an architect, particularly on smaller projects. He may be required to work from sketches prepared by the property owner, or he may proceed on the basis of verbal discussions. In such situations, the electrician operates as partial planner. Sometimes, an electrician is instructed to install a wiring system as he thinks best. In turn, he is solely responsible for system planning. Therefore, the apprentice electrician should study to achieve competence in this area, as well as in the mechanics of installation.

Lighting methods for building interiors can be classified as direct, indirect, and semi-direct arrangements. An optimum lighting system involves a judicious balance between direct and indirect illumination, as was noted previously. Observe also that every lighting installation entails the serving of specific purposes. Thus, in the home, lighting methods have the function of facilitating domestic activities and of making the interior attractive and pleasing. Again, in a store, illumination is planned to promote the sale of merchandise. Or, in a factory, lighting arrangements are planned to optimize the illumination of work on machines or in assembly positions. In business offices, lighting systems are designed to facilitate clerical work and to minimize fatigue of office personnel. Exterior lighting in patios, malls, and garden walkways should serve both decorative and utilitarian functions.

Residential interior lighting is fundamentally planned for reading, cooking, sewing, recreational activities, cleaning, and so on, with more or less attention to decorative values. Figure 13-23 depicts the basic distinction

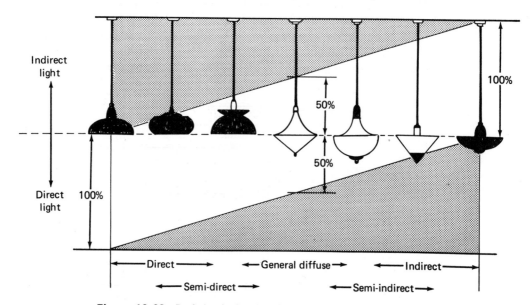

Figure 13-23 Basic luminaires ranging from direct to indirect lighting units

Figure 13-24   A small crystal chandelier

between direct and indirect lighting. These elementary luminaires (fixtures) will provide a full range from direct to indirect illumination. On the other hand, note that their decorative or esthetic value is mediocre or poor. By way of contrast, a crystal chandelier, such as illustrated in Figure 13-24, is artistically designed and provides high decorative value. Since most chandeliers provide a preponderance of indirect illumination, they are good sources of general illumination. In turn, a chandelier needs to be supplemented by local sources of diffuse direct illumination, such as floor lamps and table lamps. Note that if a floor lamp or table lamp is located away from a wall outlet, a floor outlet should be installed.

When a room has low ceilings, flush or semi-flush ceiling fixtures, as shown in Figure 13-25, provide a decorative source of general illumination. Distributed sources of general illumination may be provided by wall bracket lamps, as shown in Figure 13-26. Pole lamps are often used for local illumination or, if placed in the corner of a room, a pole lamp can provide supple-

Figure 13-25   Example of a semi-flush ceiling fixture

mentary general illumination. As a rule of thumb, the most elaborate luminaires are installed in living rooms and dining rooms. Decorative but less elaborate fixtures are installed in entrances, bedrooms, and halls. Fixtures used in kitchens and laundries tend to be useful rather than decorative.

Residential exterior lighting installations serve both utilitarian and

Figure 13-26 A wall bracket lamp provides supplementary general illumination

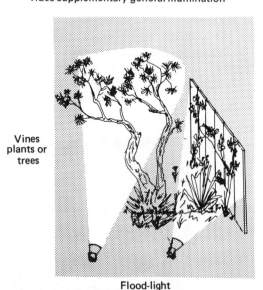

Figure 13-27 Landscaping with floodlighting

decorative functions. A driveway light, for example, might consist of a flood-bulb cluster that provides ample illumination with little decorative value. On the other hand, a driveway might be illuminated by post-top lanterns to obtain high decorative value. Mushroom and path lights provide decorative illumination for garden walkways. For contouring trees, climbing vines, or tall plants, white or colored floodlights may be installed, as shown in Figure 13-27. Note that post-top lanterns, garden lights, and path lights should be installed with underground wiring.

13–6 RETAIL STORE ILLUMINATION

Display-window lighting is planned to draw attention to merchandise and to arouse interest. Both general and direct illumination are usually required, with the light source hidden from the viewer, insofar as possible. General illumination is ordinarily provided by overhead fluorescent lamps.

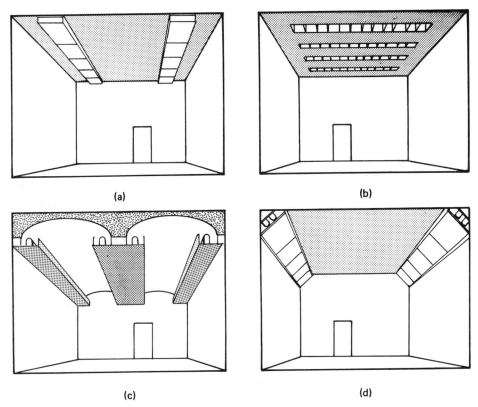

(a)

(b)

(c)

(d)

Figure 13-28 Examples of built-in luminous architectural elements (a) overhead panels (b) troffer arrays (c) cove arrangement (d) cornice lighting

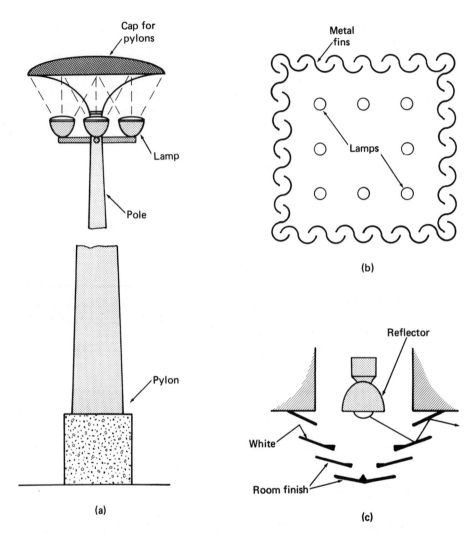

Figure 13-29 Typical built-in lighting installations with louvers (a) pylon structure (b) column cross-section (c) pilaster cross-section

Direct lighting is most effective when it originates from the direction of the viewer's line of sight. Modeling and texture can be emphasized by installing light sources at the sides of the display. Lighting from the bottom of a display window is seldom advisable, because an unnatural effect is likely to result. Light from the rear of a display window should be carefully planned. Although it can add depth to the arrangement, its potential benefit can be realized only if the light is adequately screened and diffused. Note that light sources in display windows need not bring out exactly true colors in

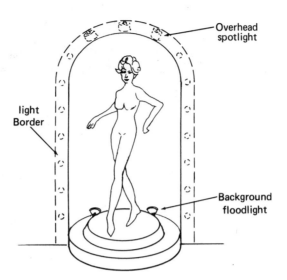

Figure 13-30 Niche illumination requires careful planning

merchandise. However, a comparatively high level of illumination is important.

Interior illumination of retail stores should bring out the true colors of merchandise. Therefore, deluxe fluorescent lamps are widely utilized, although their efficiency is less than that of ordinary fluorescent lamps. Incandescent lamps are also suitable for interior illumination. Attention to particular merchandise can be directed by built-in lamps in counters or shelves. Spotlights may be utilized to make selected items stand out prominently. In large stores, built-in luminous architectural elements with fluorescent lamps may be employed for general illumination. As an illustration, panels, troffers, coves, and cornices may be exploited in overhead installations, as shown in Figure 13-28. Other suitable elements include recesses, cavities, niches, pylons, columns, pilasters, parapets, spandrels, beams, coffers, and moldings.

An example of a pylon lighting installation is depicted in Figure 13-29(a). The pylon may be constructed of translucent material, with the cap illuminated with white or colored lamps. Additional lamps may be utilized along the pole inside the pylon. Illuminated columns may be constructed from opaque louvers, as shown in Figure 13-29(b). Illuminated pilasters may also utilize louvers for light diffusion, as shown in Figure 13-29(c). Large niches can be effectively illuminated, as seen in Figure 13-30. Background floodlighting adds a dramatic effect, when judiciously installed. Overhead spotlighting is needed for general illumination. Border lighting provides a finishing touch to the installation and increases its attention

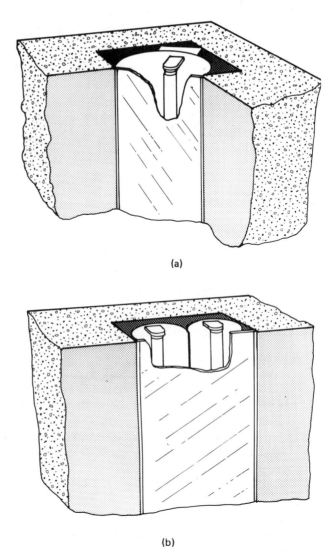

(a)

(b)

Figure 13-31 Examples of built-in lighting (a) corner panel
(b) flush panel

value. In unusually luxurious establishments, some of the general illumination may be provided by diffused green lighting beneath transparent floors. This form of illumination contributes to a feeling of "walking on clouds."

Other types of built-in lighting installations include flush-panel and corner-panel arrangements, as shown in Figure 13-31. Either white or colored lamps may be used or variegated colored lamps. Panels are made of translucent plastic in typical installations. An example of soffit lighting

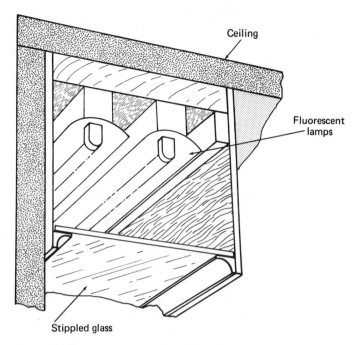

Figure 13-32 An example of soffit lighting

is depicted in Figure 13-32. Soffits are often installed over windows. Crystal glass with fine stippling may be used for paneling. Aside from the function of illumination, numerous illuminated signs are commonly installed in the larger retail stores. They indicate the location of elevators, escalators, and stairways, identify various merchandise areas, and provide specific information where needed.

13–7 OFFICE ILLUMINATION

Office illumination is utilitarian in function with more or less emphasis on decorative values. In private offices, prestige is often a dominant consideration, and lighting installations are planned in much the same manner as for luxurious homes. From the utilitarian viewpoint, clerical work is most demanding. Ample illumination free from glare is essential. Most workers are right-handed, so desks are placed with the source of daylight to the left. Similarly, recessed luminaires or fluorescent fixtures should be installed to provide general illumination from the left-hand side of the desks. Fluorescent luminaires of troffer design, such as shown in Figure 13-33(a), are well suited for office installations. Flushmounted units, such as shown in Figure 13-33(b), are also a good choice, provided the glass panels give considerable diffusion.

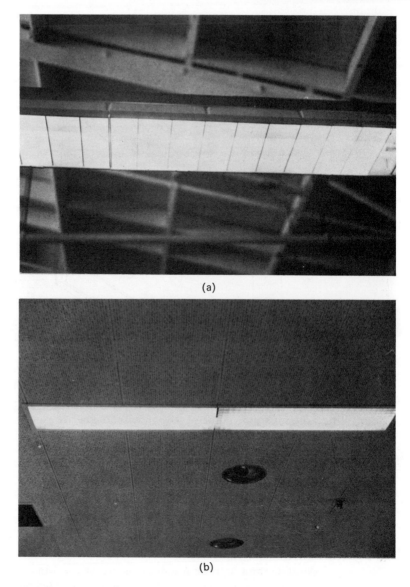

Figure 13-33 Fluorescent luminaires for office installations (a) troffer type (b) flush type

General illumination may need to be supplemented by local illumination at business machines. Desks may also require local illumination. Small fluorescent lamps are generally preferred. This type of lamp can be moved at will to an optimum position for the task at hand. Floor outlets should be installed for connection of lamps into the wiring system. Office supply rooms are sometimes neglected when planning a lighting system. Ample illumination should be available, so that time is not wasted in attempting to read small print on labels. Employee morale is also boosted by adequate lighting in all work areas, even though a particular room or station is infrequently used.

13–8 INDUSTRIAL ILLUMINATION

Industrial illumination installations are essentially utilitarian. General considerations that apply to machine and assembly-position lighting have been noted above for business-machine and office-desk lighting. In addition to fluorescent lamps, factories often utilize color-corrected mercury lamps. Mercury lamps are more economical to install and to maintain when general illumination is required over a large area. Note that the efficiency of mercury lamps is approximately the same as that of fluorescent lamps, but they are much longer lived and smaller in size. When industrial plants are modernized, conventional incandescent lamps are often replaced by mercury lamps. Note that ballast starting equipment must be installed with a mercury lamp.

Comparatively high levels of illumination should be planned for industrial installations to optimize productivity, to improve the quality of production, and to minimize the possibility of accidents. Adequate illumination throughout a factory also improves employee morale and contributes to efficient operation. Maximum illumination levels are required for detailed and close work, and at inspection stations where small flaws must be detected in manufactured products. In most factories, interior illumination is required during the brightest summer days. There is a marked trend to the building of "no daylight" factories, in which the only illumination consists of electric lamps.

13–9 SCHOOL ILLUMINATION

Classroom lighting installations are much the same as for business offices, except that supplementary illumination is seldom required. General illumination is maintained at a sufficiently high level that desks do not require local illumination. However, school workshops, laboratories, and vocational areas often require both general and supplementary lighting sources. In turn, the same planning considerations apply that have been explained for commercial and industrial installations. In the past, daylight has been exploited in school-building design for general illumination. However, vandalism in some urban and rural localities has made the use of glass windows prohibitively expensive. For this reason, there is a definite trend to the construction of "no daylight" schools, comparable with "no daylight" factories.

13–10 CHURCH ILLUMINATION

General illumination in churches is typically provided by directed ceiling fixtures pointed toward the altar. Illumination levels are comparable with residential installations. Chandeliers may be provided in larger churches for additional general illumination. Supplementary lighting for the altar and sanctuary is often provided by concealed floodlights. Outlets should be installed for pulpit lights, choir-loft illumination, and outdoor lighting for activities and displays during festival seasons. Outdoor illumination is planned in the same manner as described previously for patio and garden lighting.

STUDENT EXERCISE

This exercise provides practice in evaluating loads corresponding to illumination requirements, and in calculating the size of feeders that are utilized. Fill in the blank spaces in the following tabulation:

A retail store is 50 × 60 ft, has 3,000 sq ft of area, and has 30 ft of show window.

Calculated Load

General lighting load:

3,000 sq ft at 3 watts per sq ft × 1.25 (NEC factor) ....11,250 watts

Show-window lighting load:

30 ft at 200 watts per ft.................................. 6,000 watts

Minimum Number of Branch Circuits (NEC):

General lighting load: 11,250/230 = ____ amps for 3-wire 115/230 volts, or 98 amps for 2-wire 115 volts. (Branch-circuit load must not exceed 80% of the branch-circuit rating.) Circuits may be as follows:

Three 30-amp, 2-wire, and one 15-amp, 2-wire circuits. Or,

Five 20-amp, 2-wire circuits. Or,

Three 20-amp, 2-wire, and three 15-amp, 2-wire circuits. Or,

Seven 15-amp, 2-wire circuits. Or,

Three 15-amp, 3-wire, and one 15-amp, 2-wire circuits.

Show-window load: 6,000/230 = ____ amps for 3-wire 115/230 volts, or 52 amps for 2-wire 115 volts. Circuits may be as follows:

Four 15-amp, 2-wire circuits. Or,

Three 20-amp, 2-wire circuits. Or,

Two 15-amp, 3-wire circuits.

Minimum Size Feeders:

For 115/230-volts 3-wire system:

Current demand: 49 + 26 = ____ amps

For 115-volt system:

Current demand = 98 + 52 = ____ amps

QUESTIONS

1. What is the class of locations that are hazardous because of the presence of combustible dust?

2. What is the class of locations that are hazardous because of the presence of ignitible fibers?

3. What is the class of locations that are hazardous because of the possibility of explosive mixtures?

4. How are the Y and delta systems similar?

5. What is the advantage of using a 277-volt lighting system when heavy lighting loads must be installed?

6. The NEC places a limit on the voltage of a lighting circuit that branches. What is that voltage?

7. Where are busways permitted?

8. How many basic types of bus ducts are there?

9. What is the maximum length of a run before an expansion joint must be used?

10. What is the maximum ampacity of a short-run feeder duct?

11. What is the basic difference between using a wire and a busway for circuits?

12. What is the name given to the unit that is used to enclose switches, overcurrent devices, or other appliances?

13. Which is the least preferred conduit in heavy-current circuitry?

14. When must an explosive-proof panelboard be installed?

15. What are the three classifications of building lighting?

16. What is the basic distinction between direct and indirect lighting?

17. What two purposes does most lighting serve?

18. What are the primary purposes of retail store lighting?

19. What are the requirements for lighting in industrial plants?

14 MOTOR INSTALLATIONS

14–1 GENERAL CONSIDERATIONS

Most of the motors used in urban or rural residences and in farm out-buildings are of the single-phase type that operate at either 120 or 240 volts. A few farms install 240-volt three-phase service for operating large motors. The distinction between three-phase service and three-wire single-phase service was explained previously. All motors are rated in horsepower to indicate the useful work they can do at full load. One horsepower is equal to 746 watts. Of course, no motor is 100% efficient. A typical 1-horsepower motor is 75% efficient at full load and consumes 1,000 watts (1 kilowatt) from the line while supplying 746 watts of load power. Note that a motor may draw up to four times its rated full-load current for a short time while it is starting. This requirement must be taken into account when calculating wire gages and installing circuit breakers or fuses. Most motors run at approximately 1740 rpm when fully loaded and at nearly 1800 rpm while idling.

Many motors are designed so that they can be connected for either 120-volt or 240-volt operation. This is printed on the nameplate. Since a motor draws only half as much current in 240-volt operation as in 120-volt operation, smaller gage wires can be installed if the motor is operated at 240 volts. Table 14-1 lists the recommended wire gages for various sizes of motors operated at 120 volts or 240 volts, with lines from 25 to 400 feet in length. Note that a motor can be overloaded for brief intervals without damage, but the motor will heat up excessively and burn out if subjected to continued overload.

It is advisable to install motor starting and stopping switches, such as depicted in Figure 14-1. This type of switch has a built-in protective element

Table 14-1 Circuit wire sizes for individual single-phase motors (Courtesy Leviton Inc.)

Horsepower of Motor	Volts	Approximate Starting Current Amperes	Approximate Full Load Current Amperes	Feet	LENGTH OF RUN IN FEET from Main Switch to Motor							
					25	50	75	100	150	200	300	400
¼	120	20	5	Wire Size	14	14	14	12	10	10	8	6
⅓	120	20	5.5	Wire Size	14	14	14	12	10	8	6	6
½	120	22	7	Wire Size	14	14	12	12	10	8	6	6
¾	120	28	9.5	Wire Size	14	12	12	10	8	6	4	4
¼	240	10	2.5	Wire Size	14	14	14	14	14	14	12	12
⅓	240	10	3	Wire Size	14	14	14	14	14	14	12	10
½	240	11	3.5	Wire Size	14	14	14	14	14	12	12	10
¾	240	14	4.7	Wire Size	14	14	14	14	14	12	10	10
1	240	16	5.5	Wire Size	14	14	14	14	14	12	10	10
1½	240	22	7.6	Wire Size	14	14	14	14	12	10	8	8
2	240	30	10	Wire Size	14	14	14	12	10	10	8	6
3	240	42	14	Wire Size	14	12	12	12	10	8	6	6
5	240	69	23	Wire Size	10	10	10	8	8	6	4	4
7½	240	100	34	Wire Size	8	8	8	8	6	4	2	2
10	240	130	43	Wire Size	6	6	6	6	4	4	2	1

Figure 14-1 A motor starting switch with built-in overload protection

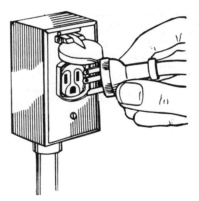

Figure 14-2 Weatherproof motor outlets are required in damp locations

so that the motor will not be damaged in case of an excessive starting overload. If an extension cord is used to connect a motor to an outlet, a suitable heavy type of cord must be selected. Note that ordinary lamp cord can never be used for motor operation. In outdoor or farm-outbuilding installations, weatherproof outlets (Figure 14-2) must be provided for plugging motor cords into the wiring system. The housing of a motor is grounded via the white wire in the line so that the operator will not receive a shock in case the motor windings break down.

14–2 MOTOR CIRCUITING

When a wiring system is to be installed for more than one motor, there are five basic arrangements that may be employed. Figure 14-3(a) shows a layout in which a separate circuit is run to each motor from a power panel-

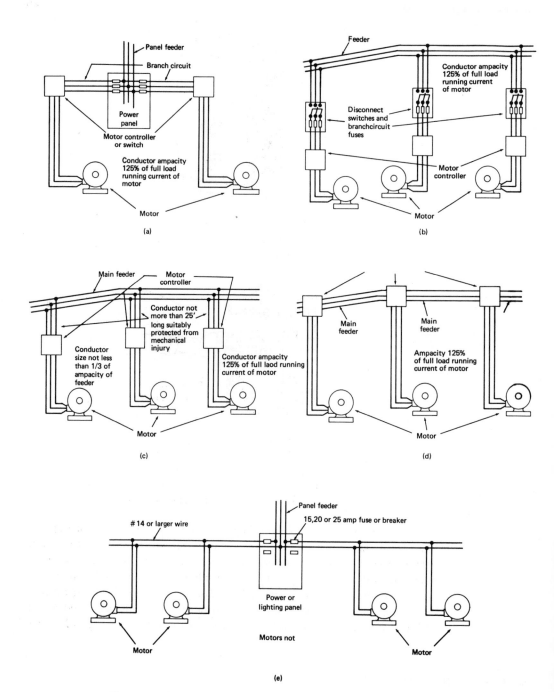

Figure 14-3 Wiring arrangements for supplying more than one motor (a) individual circuit to each motor from a power panelboard (b) branch circuits tapped from feeder at various points (c) branch circuits with fused feeder only (d) feeders connected directly to motor controllers (e) small motors paralleled on a branch circuit

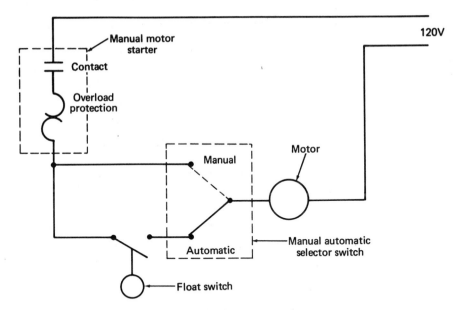

120V

Figure 14-4 Example of manual/automatic motor switch operating a sump pump

board. Note that the conductor ampacity must be at least 125% of the full-load running current. If the distance from the power panelboard to the motor is substantial, the conductor ampacity may need to be increased in order to avoid excessive voltage drop. Figure 14-3(b) shows an arrangement in which a feeder is tapped by branch circuits at various points. That is, no power panelboard is used as a distribution center for the various branch circuits.

Branch-circuit fuses are omitted in the arrangement depicted in Figure 14-3(c). Overload protection for the individual motors may be provided in the associated controller, but some motors have built-in thermal overload protection devices. Note that the National Electrical Code limits the length of conductors from the main feeder to the motor controller to not more than 25 feet when this arrangement is utilized. Sometimes it is practical to connect the main feeder directly to the motor controllers, as shown in Figure 14-3(d). If small motors that do not draw over 6 amperes each at full load are installed, it is permissible to parallel two or more motors on a branch circuit, as depicted in Figure 14-3(e).

Some motors are switched on and off automatically, as well as manually. For example, an air-conditioner motor can be switched on manually and its operating intervals thereafter are controlled by a thermostat. The same principle is employed in refrigerator motor circuits, sump pump circuits, and so on. Figure 14-4 shows a motor circuit for a sump pump. Note that thermal overload protectors may be either of the automatic or the manual reset type. In the case of an automatic overload protector, the circuit will open when the current demand exceeds a preset value for a certain length

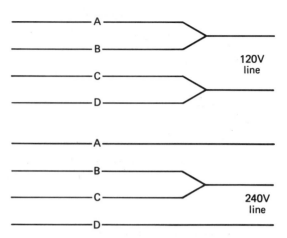

Figure 14-5 Terminal connections of motor for 120-volt and 240-volt operation

of time. Then, after a certain waiting period, the protector will close the circuit once more. On the other hand, in the case of a manual reset protector, a button is tripped if a substantial overload occurs. In turn, the circuit remains open until the reset button is pressed.

A 120-volt motor requires one fuse in the "hot" side of the line. That is, the neutral side of the line is never fused because a shock hazard would exist if the neutral circuit were opened. A 240-volt motor requires two fuses, with one fuse in each of the "hot" lines. However, if a 240-volt three-phase motor is installed, three fuses are required, with one fuse in each of the "hot" lines, as seen in Figures 14-3(a) and (b). Three-phase motor circuits are explained in greater detail subsequently. In the example of Figure 14-3(e), a 120/240-volt feeder is supplying two 120-volt motors in parallel. One fuse is installed in the "hot" line. Time-delay fuses with a rating of 175% of the motor full-load current demand are commonly installed to provide for starting-current demands. Or, if a circuit breaker is employed, it may have a rating of 250% of the motor full-load current demand.

Figure 14-5 shows how the terminals of a 120/240-volt motor are connected for 120-volt and for 240-volt operation. As noted above, the 120-volt connection requires one fuse, whereas the 240-volt connection requires two fuses. In rural installations where a grounding wire may not be available at the motor, the housing of the motor must be grounded by a separate conductor runing to a cold-water pipe or to a ground rod. Note that it is permissible to operate an ungrounded portable motor, provided the circuit has a potential difference of less than 150 volts to ground. Nevertheless, it is recommended that such motors be operated with a three-wire cord, so that the housing can be grounded and shock hazard avoided.

14–3 SYNCHRONOUS MOTORS

All of the AC motors that have been discussed previously are induction motors. An induction motor does not have constant speed under load—as the load is increased, the motor runs slower. This reduction in speed is called slip. There is a small amount of slip in an induction motor even under no load. For example, a motor that has a theoretical no-load speed of 1,800 rpm will have an actual no-load speed of approximately 1760 rpm. This slip factor is of no practical concern in power applications such as refrigerators, washing machines, electric drills, and so on. However, an induction motor cannot be used in applications that require exact speed of rotation. As an illustration, an induction motor cannot be used in an electric clock. If a clock is to keep exact time, its motor must rotate at an exact speed.

Electric clocks utilize the synchronous type of AC motor. A synchronous motor has no slip, and it rotates in exact step with the power-line frequency. A synchronous motor is somewhat more elaborate and costly than an induction motor. When electric clocks were first manufactured, they were not self-starting. In other words, after the clock had been connected to a power outlet, the motor had to be given a "start" with the fingers before it would rotate and lock into synchronism with the 60-Hz line frequency. This type of electric clock could be made to run backward if the motor was given a "start" in the reverse direction. However, later electric clocks were made self-starting by the addition of small coils called shading coils. A shading coil splits a single-phase AC source into a two-phase source. Note that a motor that is started by means of shading coils cannot be made to run backward. In other words, if the motor is given a "start" in the reverse direction, this rotation is opposed by the action of the shading coils. Or, if the plug is reversed in the power outlet, the current through the motor coil is reversed and the current through the shading coils is also reversed. In turn, the motor automatically starts in the correct direction of rotation.

STUDENT EXERCISE

This exercise provides practical experience with the line drops that occur during a motor-starting interval, and in normal running operation.

1.   Obtain a volt-amp-wattmeter and make the experiment with a refrigerator or a freezer.

2.   Plug the instrument cord into a 120-volt outlet, and plug the refrigerator cord into the receptacle provided by the instrument.

3.   Open the door of the refrigerator, so that the thermostat will become actuated.

4.   Note the line voltage before the thermostat closes the motor circuit.

5.   As soon as the motor circuit is closed, note the lowest dip that occurs in the line voltage.

6.   After the motor speeds up to normal operation, note the rise in line voltage.

7.   Calculate the percent voltage drop that occurred during the starting interval and during the running interval.

8.   Repeat the experiment and measure the current demand of the refrigerator motor during the starting interval and the running interval.

9.   Apply Ohm's law and calculate the resistance of the branch line that supplies the motor.

STUDENT EXERCISE

This exercise provides practical experience with the use of current shunts used by motor electricians.

1.   Obtain a current shunt such as illustrated in Figure 14-6 and a matching AC ammeter. The ammeter should have a top range of at least 15 amperes.

2.   This experiment is made with a refrigerator. Connect the current input terminals of the shunt in series with one side of the line. Connect the ammeter to the meter terminals on the shunt.

3.   Plug the cord with its shunt and meter into a 120-volt outlet.

4.   Open the door of the refrigerator, so that the thermostat will become actuated.

5.   As soon as the motor circuit is closed, note the peak current demand.

6.   After the motor speeds up to normal operation, note the running current value.

7.   Calculate how many times more starting current is demanded, than running current.

8.   Apply Ohm's law and calculate the starting impedance and the running impedance values of the motor, using the line-voltage values obtained in the previous exercise.

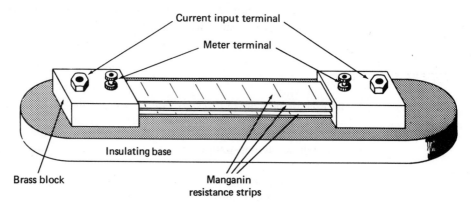

Figure 14-6   A current shunt for an ammeter

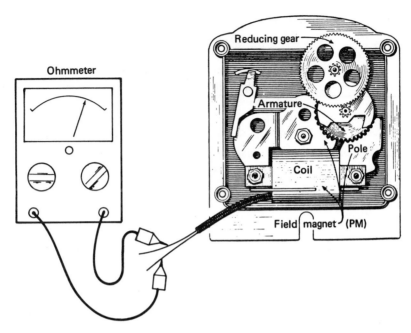

Figure 14-7   An electric-clock motor

STUDENT EXERCISE

This exercise provides familiarity and practical experience with very small motors.

1. Obtain an electric clock such as depicted in Figure 14-7 and an ohmmeter.
2. Connect the ohmmeter test leads across the coil terminals.
3. Observe the meter reading; a typical resistance indication is 800 ohms.
4. Calculate the value of current that would be drawn from a 120-volt DC source.
5. Will the coil draw more or less current from a 120-volt AC source? Why?
6. Would the clock operate if connected to a 120-volt DC source? Why?
7. Connect the clock to a 120-volt AC source, and stop the motor with your fingers to observe the amount of power that is developed.
8. Reverse the plug in the power outlet, to determine whether the motor now runs backward.
9. Can you explain your observation in step No. 8?

QUESTIONS AND EXERCISES

1. How many watts are equal to 1 horsepower?
2. What is the typical efficiency of an electrical motor?

3. What is a typical ratio of starting to running current for an electrical motor?

4. What is the National Electrical Code's limit to the length of the main feeder to the motor controller in a setup such as illustrated in Figure 14-3(c)?

5. When is it permissible to parallel two or more motors on a branch circuit?

6. What is the difference between a manual reset protection and a thermal overload protection?

7. Why is the neutral line to a motor never fused?

8. Why are time-delay fuses usually used in motor circuits?

9. Draw a diagram showing the correct method for connecting a motor for both 120-volt and 240-volt operation.

15 AUTOMOTIVE WIRING

15–1 GENERAL CONSIDERATIONS

Automotive wiring is originally installed by factory electricians. If the wiring system becomes defective, it is repaired by a garage electrician or an automotive electrical specialist. An original wiring installation sometimes requires reworking because of the addition of various electrical accessories. That is, an original circuit may become overloaded or a new accessory may require modification and/or elaboration of an original circuit. Sometimes one type of vehicle is converted into another type, with the result that extensive reworking of the original wiring system is required. As an illustration, a pickup may be converted into a camper, or a van may be converted into a mobile shop.

Regardless of the wiring system that is utilized, the greatest current demand is imposed by the starting motor, particularly during cold weather. The starting-motor circuit is generally wired with No. 1 or No. 0 gauge stranded copper conductor. As noted previously, the outer diameter of an insulated cable is not a reliable indication of the conductor gauge. That is, a replacement cable should be checked specifically for the gauge of its conductor. The thickness of the insulation and the lay of the cable both contribute to its outer diameter. A concentric lay has approximately 25% more copper than a rope lay of strands for a given diameter of conductor. It is advisable to select a cable that is marked for conductor size.

Two types of cables are used in primary circuits of ignition coils. One type is made of conventional stranded copper wire and the other type consists of a resistance wire incorporated in the instrument-panel wiring harness. It has a lower resistance when cold than when the engine comes up to operating

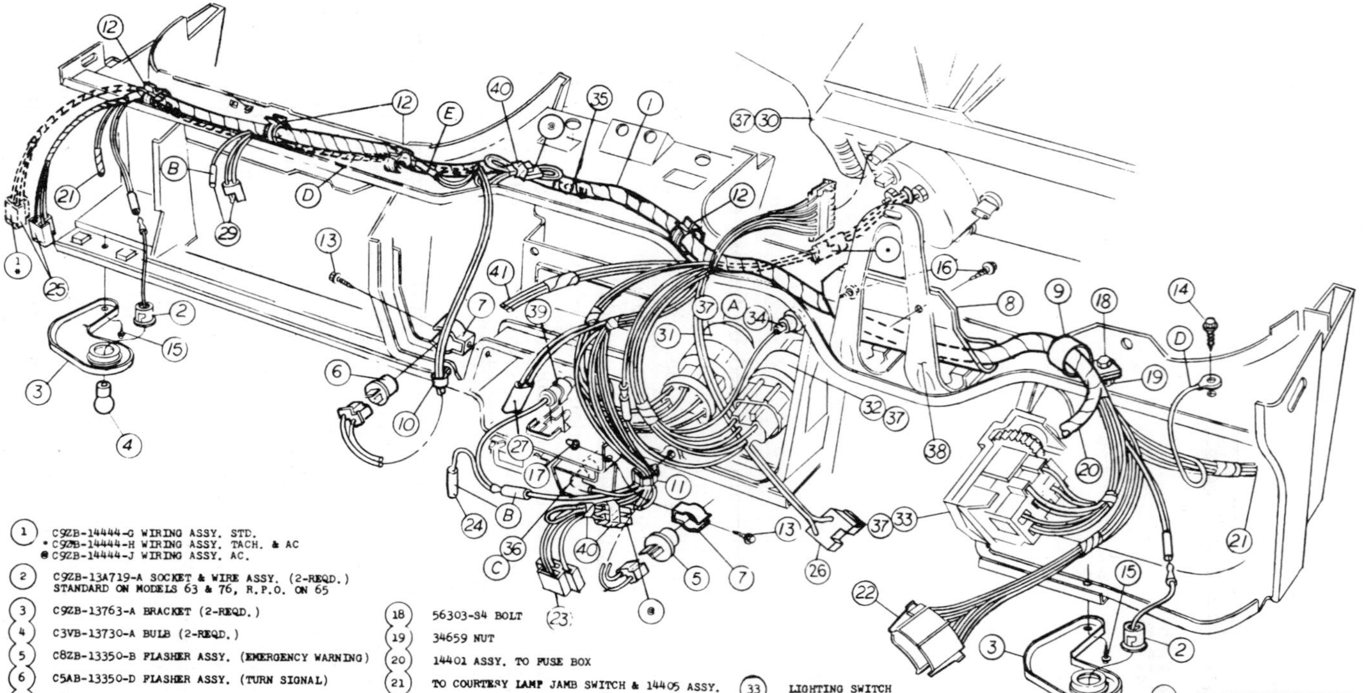

Figure 15-1 A typical instrument panel wiring harness (Courtesy Ford Motor Company)

1. C9ZB-14444-O WIRING ASSY. STD.
   C9ZB-14444-H WIRING ASSY. TACH. & AC
   C9ZB-14444-J WIRING ASSY. AC.
2. C9ZB-13A719-A SOCKET & WIRE ASSY. (2-REQD.)
   STANDARD ON MODELS 63 & 76, R.P.O. ON 65
3. C8ZB-13763-A BRACKET (2-REQD.)
4. C3VB-13730-A BULB (2-REQD.)
5. C8ZB-13350-B FLASHER ASSY. (EMERGENCY WARNING)
6. C5AB-13350-D FLASHER ASSY. (TURN SIGNAL)
7. C8ZB-13763-A BRACKET (2-REQD.)
8. C9ZB-14A099-E SHIELD
9. 378674-S CLIP
10. 382929-S CLIP (2-REQD.)
11. 382930-S CLIP (2-REQD.)
12. 383207 CLIP (4-REQD.)
13. 55906-S36 SCREW (2-REQD.)
14. 55907-S36 SCREW
15. 55927-S36 SCREW (2-REQD.)
16. 40927-S2 SCREW (2-REQD.)
17. 353538-S STUD PROTECTOR
18. 56303-S4 BOLT
19. 34659 NUT
20. 14401 ASSY. TO FUSE BOX
21. TO COURTESY LAMP JAMB SWITCH & 14A05 ASSY.
22. TO STEERING COLUMN
23. TO HEATER BLOWER SWITCH
24. TO HEATER ILLUMINATION
25. TO HEATER RESISTOR
26. TO STOP LAMP SWITCH
27. TO RADIO RECEIVER
28.
29. TO CLOCK & MAP LAMP
30. INSTRUMENT CLUSTER
31. WINDSHIELD WASHER SWITCH
32. IGNITION SWITCH
33. LIGHTING SWITCH
34. IGNITION SWITCH ILLUMINATION
35. POSITION LOCATOR IN HOLE PROVIDED
36. CIGAR LIGHTER
37. FOR INSTALLATION OF CONTROLS & INSTRUMENT CLUSTER
38. BRACK ASSY. BRAKE PEDAL SUPPORT
39. C9AB-13A726-A LAMP ASSY. (ASH TRAY) (R.P.O.) ONLY
40. E3B-M36-71 3/4" TAPE TIE CONN. WHEN A/C IS NOT SPECIFIED
41. ENGINE COMPARTMENT WIRING

A. 11A BLACK YELLOW-STRIPE
B. 19D, E BLUE RED-STRIPE
C. 40 BLUE WHITE-STRIPE
D. 57 BLACK
E. 54 GREEN-YELLOW STRIPE

NOTE: *FOR R.P.O. CLUSTER ASSY. WITH TACH. & A/C
@FOR A/C WIRING

324

temperature. It is essential to install the correct type of primary cable for a particular ignition system. Figure 15-1 shows the appearance of a typical instrument-panel wiring harness. Two types of high-tension ignition cables are in general use. Both types have heavy insulation of rubber and plastic to withstand the high voltage applied to the spark plugs. The first type employs conventional copper or aluminum stranded conductors, while the other type has a fiberglass core that is impregnated with carbon. This impregnation provides a resistance that minimizes interference to radio and television receivers.

15–2 SCHEMATIC SYMBOLS

Schematic symbols used in automotive electricity are different in some cases from the symbols utilized in power wiring diagrams and architectural wiring diagrams. Figure 15-2 depicts the circuit and switch symbols employed

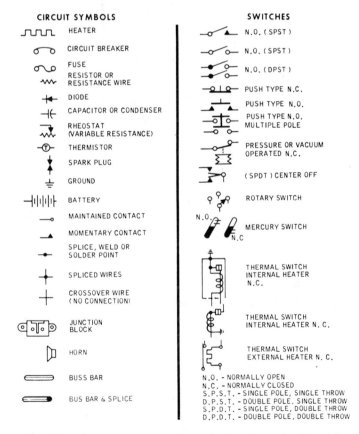

Figure 15-2 Circuit and switch symbols used in automotive wiring diagrams (Courtesy Ford Motor Company)

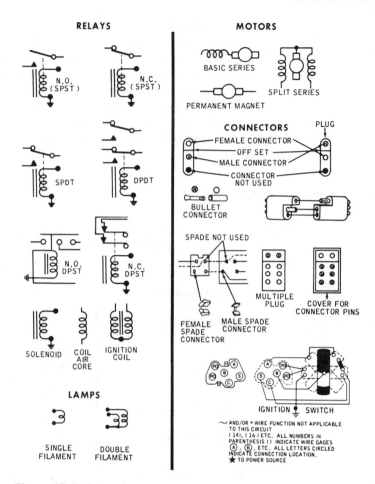

Figure 15-3 Relay, lamp, and motor symbols used in automotive wiring diagrams (Courtesy Ford Motor Company)

in automotive wiring diagrams. Note in passing that one terminal of the storage battery is always grounded to the framework of the body when installed in an automobile. Most wiring systems use a negative ground; however, some systems use a positive ground. A heater element in an automotive circuit is not symbolized as a resistance, but as a rectangular form. A spark plug is symbolized by a pair of proximate contacts. Bus bars and bus bars with splices have specialized symbols. An SPDT switch with a center-off position is denoted by a symbol that differs from that used in power diagrams.

Figure 15-3 shows the relay, lamp, and motor symbols used in automotive wiring diagrams. Relay and coil symbols are essentially the same as in power diagrams. Lamp symbols are modified to some extent in that a circular outline is omitted, with the result that the remaining filament symbol appears

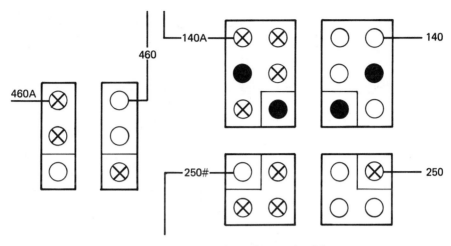

Figure 15-4 Most wire connectors are shown in open-book form

similar to a coil symbol. Male and female connector terminals are symbolized by small circles; the former encloses an X while the latter is blank. The symbol for an ignition switch is unique as it has no counterpart in other fields. It has, Off, Accessory-only, Accessory-and-coil, and Coil-start-and-proveout positions. In this last position, the starting-motor relay is energized. Connectors for wires are generally shown in open-book form, as illustrated in Figure 15-4. Note also that all "live" terminals have corresponding numbers, such as 140 and 140A, 250 and 250#, etc.

15-3 SCHEMATIC AND PICTORIAL DIAGRAMS

Both schematic and pictorial wiring diagrams are needed when automotive wiring is installed. A schematic diagram shows the circuits and connection details of the wiring system. On the other hand, schematics do not show how harnesses are routed. A pictorial wiring diagram indicates how the harnesses are run through the various spaces of the chassis, but does not identify individual wires and the terminals to which they are connected. Figure 15-5 is a schematic diagram for an exterior-light, turn-signal, and brake lights. Each conductor is identified by number, and each connection point is detailed. Devices are not necessarily shown in correct relative locations. The basic purpose of a schematic diagram is to clearly designate the connections of device and harness terminals.

To trace a circuit, it is advisable to start at a ground point, tracing it through all connectors to the source and back to ground. Note that relays and switches are shown in the "system off" position and that each wire in a harness is color-coded for ready identification. Note that manufacturing specifications for harnesses are occasionally modified during a particular pro-

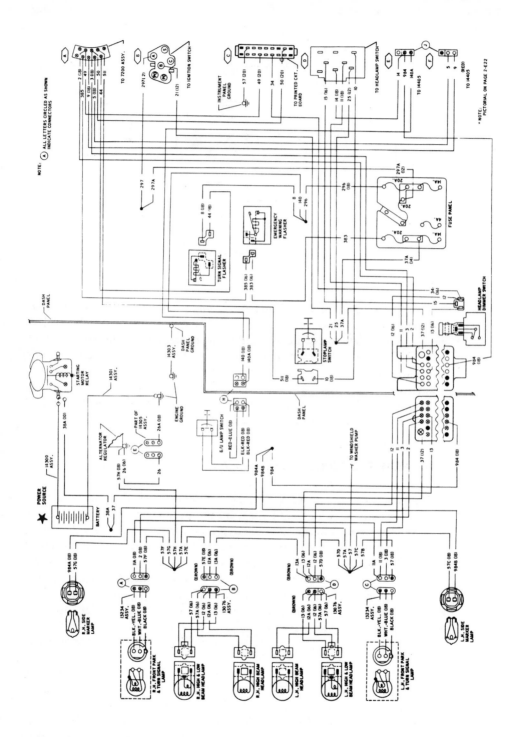

Figure 15-5 Schematic diagram for an exterior-light, turn-signal, and horn installation (Courtesy Ford Motor Company)

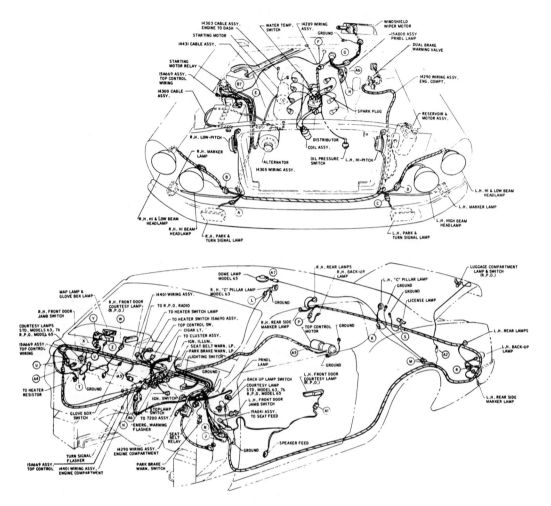

Figure 15-6 Pictorial wiring diagram for an exterior-light, turn-signal, and horn installation (Courtesy Ford Motor Company)

duction run. For this reason, the electrician might encounter a specific wire color in a connector that does not coincide with the color called for by the pictorial wiring diagram. In such a case, the wire can be identified by elimination; that is, the other colors called for at the connector will be correct. Color deviations of this sort in a particular production run are usually for only a short duration.

The routing of harnesses is planned by the manufacturer to minimize the possibility of abrasion which could eventually cut through the insulation and cause a short circuit. As seen in Figure 15-5, at bottom right, a fuse panel is provided and a fuse would blow in the event of an accidentally grounded conductor. When repairing automotive wiring it is sometimes a temptation to shorten a conductor to restore a broken-off lug, for example. This is poor practice because the shortened conductor or harness may then be positioned incorrectly and subjected to excessive tension and progressive abrasion. Instead of shortening a standard conductor or harness, it is advisable to install an exact replacement. Insulation can be damaged by excessive heat or by traces of battery acid, in addition to mechanical forces.

One schematic diagram may not be entirely complete, particularly with respect to electrical sources. For example, in Figure 15-5, a star is shown in the top left-center of the diagram. This position is noted "to power source." The battery is not explicitly indicated in this diagram. However, the schematic diagram for the ignition and starting installation (Figure 15-7) shows the battery. Note that this is a negative-ground installation. It would be a serious error to connect a replacement battery into the circuit with incorrect polarity. Two distributors are depicted in Figure 15-7, because this diagram is intended to cover both six-cylinder and eight-cylinder models. In turn, the electrician merely disgards the distributor circuitry which does not apply to the particular job.

A pictorial wiring diagram for the eight-cylinder model is shown in Figure 15-8. Cable clamps are indicated at strategic points to minimize the possibility of abrasion to the cables. It is poor practice to omit a clamp, just as it is poor practice to route a conductor otherwise than specified in the diagram. There are occasions in practice when a factory wiring diagram is unavailable for a particular accessory. For example, if a motorized radio antenna is not supplied as standard equipment and the electrician is called upon to install a motorized antenna, he will not be able to work from a pictorial diagram. In situations of this kind an electrician must take advantage of his experience and use his own best judgment.

Figure 15-7 shows how notes and dotted-line conductors are employed in schematic diagrams that apply to more than one model of an automobile. Thus, the engine-temperature switch and the oil-pressure switch shown in the lower left-hand corner are used in the Fairlane model only. In turn, conductors 642, 39, and 31 are shown dotted. Separate instrument-panel clusters are depicted for the Fairlane, Falcon, and Montego models. Note that dotted-line conductors 31 and 39 continue past the second connector as solid lines to the temperature switch and oil-pressure switch used in the Falcon and Montego models. When a conductor of a harness is not used in a wiring system, it is called a blind circuit. As an illustration, conductor 48 is a blind circuit that terminates in the harness. Note that a large black dot in a connector denotes a blank (unused) terminal. On the other hand, a large black dot along a conductor denotes a splice. Thus, conductors 37, 37A, and 25 are spliced to conductor 21 at the large black dot.

Figure 15-9 shows a schematic wiring diagram for a courtesy-light installation, and Figure 15-6 shows the corresponding pictorial wiring diagram.

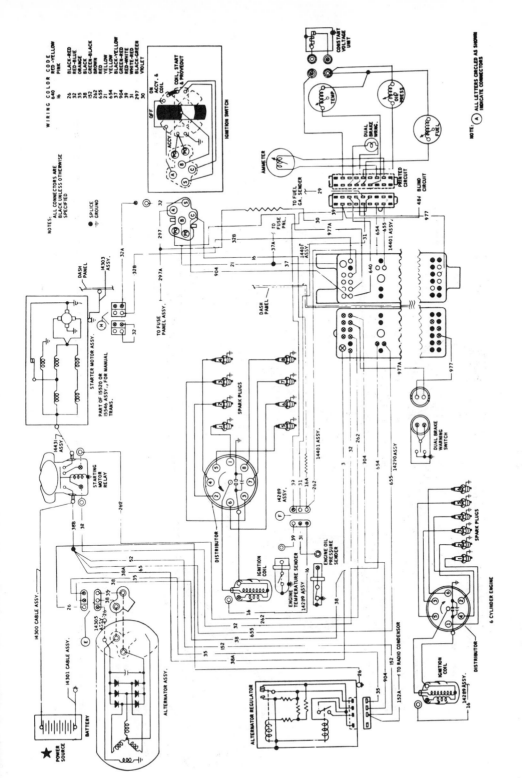

Figure 15-7 Schematic diagram of a typical ignition, charging, gages, and starting installation (Courtesy Ford Motor Company)

331

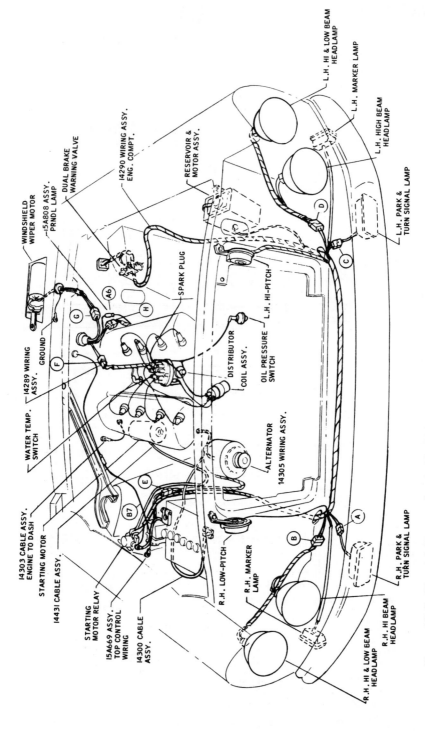

WINDSHIELD
WIPER MOTOR

15A808 ASSY.
PRNDL LAMP

DUAL BRAKE
WARNING VALVE

14290 WIRING ASSY.
ENG. COMPT.

RESERVOIR &
MOTOR ASSY.

L.H. HI & LOW BEAM
HEADLAMP

L.H. MARKER LAMP

L.H. HIGH BEAM
HEADLAMP

L.H. PARK &
TURN SIGNAL LAMP

SPARK PLUG

L.H. HI-PITCH

DISTRIBUTOR

COIL ASSY.

OIL PRESSURE
SWITCH

ALTERNATOR
14305 WIRING ASSY.

14289 WIRING
ASSY.

GROUND

WATER TEMP.
SWITCH

14303 CABLE ASSY.
ENGINE TO DASH

STARTING MOTOR

14431 CABLE ASSY.

STARTING
MOTOR RELAY

15A669 ASSY.
TOP CONTROL
WIRING

14300 CABLE
ASSY.

R.H. HI & LOW BEAM
HEADLAMP

R.H. HI BEAM
HEADLAMP

R.H. MARKER
LAMP

R.H. LOW-PITCH

R.H. PARK &
TURN SIGNAL LAMP

Figure 15-8 Pictorial diagram of typical ignition, charging, gages, and starting installation (Courtesy Ford Motor Company)

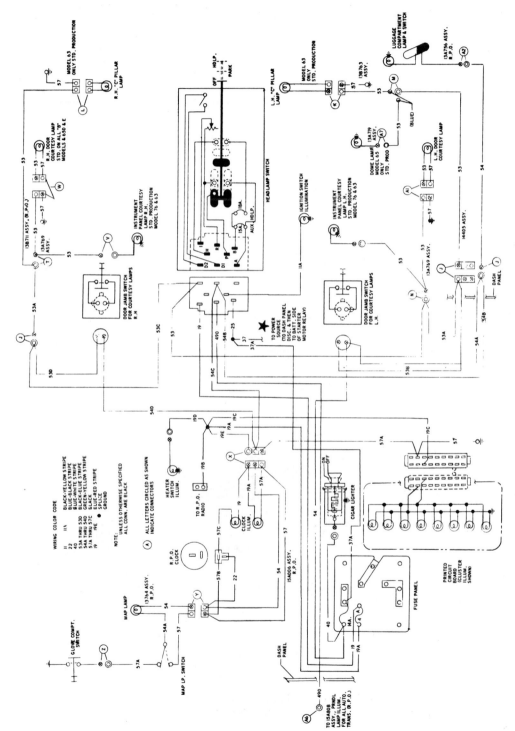

Figure 15-9 Schematic wiring diagram for interior lights, clock, and cigar lighter installation (Courtesy Ford Motor Company)

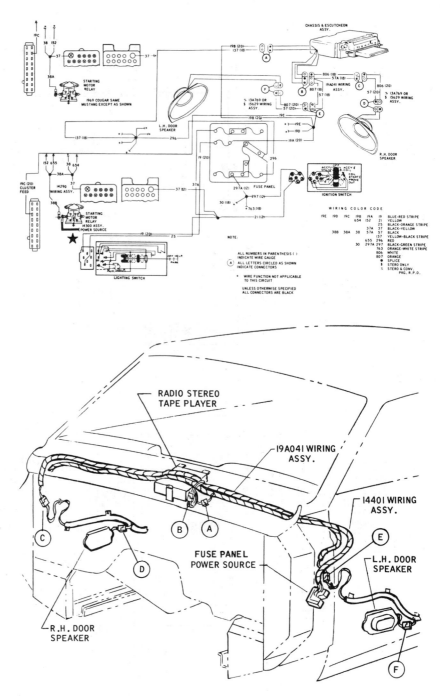

Figure 15-10 Schematic and pictorial wiring diagrams for a typical AM or AM-FM radio and speaker installation (Courtesy Ford Motor Company)

Schematic and pictorial wiring diagrams for an AM or AM-FM radio and speaker installation are shown in Figure 15-10. The technical features in these diagrams are the same as explained previously, and they serve to illustrate the appreciable complexity of an automotive wiring system. In addition to the accessories that have been described, electricians are occasionally called upon to install devices such as automatic headlight dimmers and automatic lighting-system switches. Such accessories are shipped with wiring diagrams and installation instructions.

STUDENT EXERCISE

This exercise provides familiarity with automobile wiring systems.

1. Check the terminal posts on the storage battery of your car, or another student's car. Is the larger post a positive terminal or a negative terminal?
2. Does the positive or negative terminal of the battery connect to the frame of the car?
3. Is the foregoing answer true of all cars and trucks?
4. What would happen if the battery terminals were accidentally reversed?
5. Observe the high-tension ignition harness. If the harness required replacement, how could you determine correct connections from the distributor cap to the spark plugs?
6. Locate the fuse panel. If a fuse is suspected of being open, how could you make a quick check without inserting a replacement fuse?
7. Disconnect a high-tension lead from a spark plug with the motor stopped, and place the end of the lead about an inch from the motor block.
8. Start the car, and observe how far the spark will jump from the disconnected lead. (Warning: Do not touch the terminal of the high-tension lead.)
9. Stop the motor and reconnect the high-tension lead to the spark plug.
10. Start the car, and observe how far the spark will jump from a spark-plug terminal to the motor block. Use an insulated screwdriver for this test.

QUESTIONS

1. What device in an automotive electrical system uses the greatest amount of current?
2. How much more copper does a concentric lay have than a rope lay of the same diameter?
3. What is the purpose of the high-tension ignition cable that uses a fiberglass core that is impregnated with carbon?
4. How is a spark plug symbolized?
5. What type of wiring diagrams are necessary for the installation of automotive wiring?

6. What is the purpose of the routing of a wiring harness?

7. How should you repair a conductor that has broken?

8. What is the name given to a conductor that is not used in a cable?

9. What is the meaning of a large black dot in a connector?

10. What is the meaning of a large black dot along a conductor?

16 MARINE WIRING

16–1 GENERAL CONSIDERATIONS

Marine electrical wiring starts as soon as the keel of a ship is laid. Temporary wiring is installed first to provide lighting and ventilating motors, and for electric welding equipment. Permanent wiring is installed progressively until the ship is completed. Figure 16-1 shows how a temporary 120-volt lighting feeder is installed with a transformer, switch panel, and lighting circuit. Various distribution centers are installed in the vicinity of gangplanks. Switch panels are mounted on temporary wooden frames. Cables are secured to overhead beams with beam clamps on which wooden wiring cleats are fastened. Weatherproof sockets with wire guards over the bulbs are spliced along the cables at approximately 8-foot intervals. Plugboards are also spliced along the cables for supplying portable lights, grinders, and drilling machines.

Electric arc welding is widely employed in shipbuilding and suitable generators are commonly installed beside the ship. Figure 16-2 depicts a typical feeder arrangement for portable arc and spot welders and other 480-volt three-phase equipment. Note that the metal housings of the distribution panels are grounded. Similarly, the hull of the ship and metal building structures are grounded. Four-conductor cable is run from the distribution panel on the ship to the feeder distribution panel on the ways. The fourth conductor serves to provide a continuous ground circuit for the installation.

16–2 SUBSTATION CIRCUITING

Welding units may also be energized by motor-generators installed in a substation near the ship, as depicted in Figure 16-3. The generators typically provide a 70-volt DC output and are operated in parallel to meet

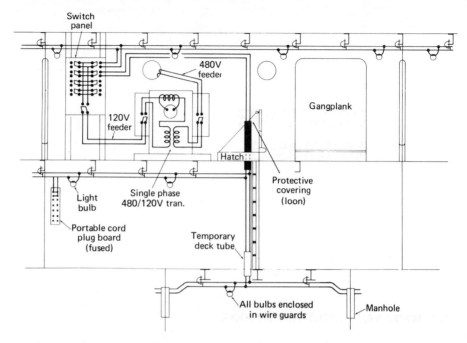

Figure 16-1 A temporary 120-volt lighting feeder with switch panel and lighting circuit

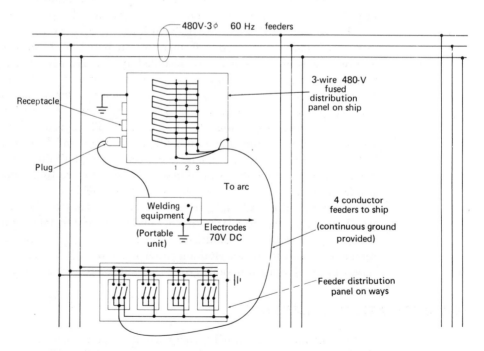

Figure 16-2 Typical feeder arrangement for 480-volt 3-phase supply to portable equipment

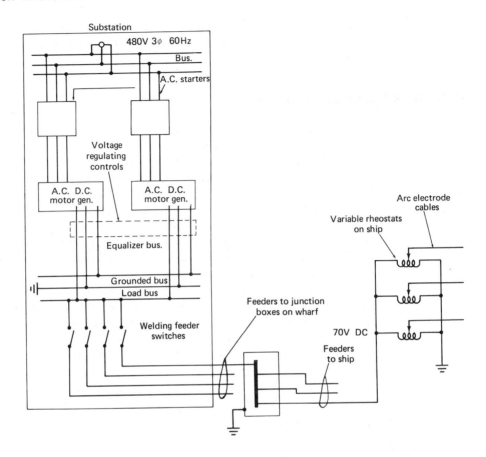

Figure 16-3 Substation circuiting for arc-welding feeders

heavy current demands. Each DC generator is grounded at one terminal; either the positive terminal or the negative terminal may be used, but the same polarity terminal must be grounded on each machine. Note that voltage controls are provided in the equalizer bus between the generators and the welding feeder switches. These controls ensure that each generator supplies exactly the same voltage to the feeders and thereby carries its proper portion of the total load.

A 70-volt arc welder may draw from 10 to 70 amperes, and a spot welder operating at 40 volts may draw 400 amperes. Substation buses, welding feeders, and cables must be chosen of sufficient ampacity to avoid appreciable voltage drop under full load. It is good practice to solder the ends of the cables into heavy lugs instead of using pressure-type connectors or screw terminals. This type of soldering requires knowledge and experience to ensure that all strands of the cable are soldered together and to the inner surface of the lug. If a connection is defective, it is likely that it will burn out under heavy load. When testing an arc-welding unit, great care must be taken to avoid getting a "flash" in the unprotected eyes, as eye injury or blindness will result.

16–3 180-HZ WIRING SYSTEM

Equipment such as portable drills, saws, grinders, and other motorized tools are lighter in weight when designed to operate from a high-cycle system instead of 60 Hz. Shipyards generally employ a 240-volt 180-Hz supply for this purpose. A frequency converter (motor-generator set) is utilized and is installed in a substation near the shipway. A cable is run from the converter to a distribution panel with half a dozen receptacles on the deck of the ship, or a number of high-cycle distribution panels may be temporarily installed. In turn, various high-cycle appliances are plugged into the distribution panel(s) as required. Figure 16-4 shows a typical high-cycle four-wire three-phase circuit with a continuous ground.

Note that a 4-wire cable is installed to provide a continuous ground from the converter to the appliance(s) in use. The ground wire in each cable is connected to the metal housings of all panels and boxes, and to the metal housings of all portable appliances or machines. As seen in Figure 16-4, the feeders are fused, but no fuses are utilized in any part of the ground cir-

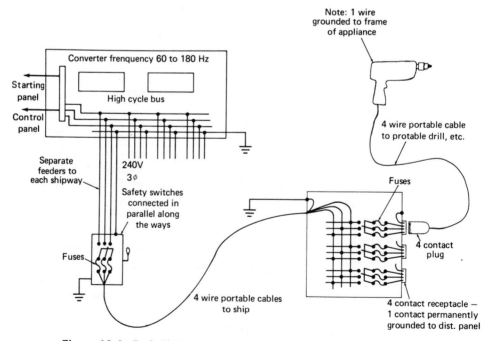

Figure 16-4 Basic high-cycle circuiting

cuit. Heavier insulation is used on 180-Hz portable cords than on 60-Hz cords. High-cycle current is more dangerous than low-frequency current. In turn, the system must be installed in accordance with good practices to minimize the danger of shock to the workmen. Moreover, work stoppage can be very costly in terms of man-hours that are lost. Therefore, temporary wiring is installed with high reliability as a prime consideration.

16–4 CABLES AND FITTINGS

Marine cables generally consist of stranded copper conductors with rubber, cloth, or asbestos insulation. From 1 to 44 strands may be utilized, with conductor cross-sectional areas from 2,000 to 1,000,000 CM. Rubber insulation is used chiefly in low-voltage circuiting. Although rubber is flexible and moisture-proof, it tends to age rapidly. Linen tape (cambric) insulation is widely used in marine installations. Cambric-insulated cables are utilized in high-voltage runs. Asbestos-insulated cables are used to a minor extent in marine installations, in locations where very high temperatures must be contended with. Lighting and power cables often contain jute fillers between the wires to provide an outer circular shape. The exterior surface may be braid, rubber, armor, or lead.

Armored cable provides protection against mechanical damage. The armor consists of metallic fabric called basket-weave. Since the metallic

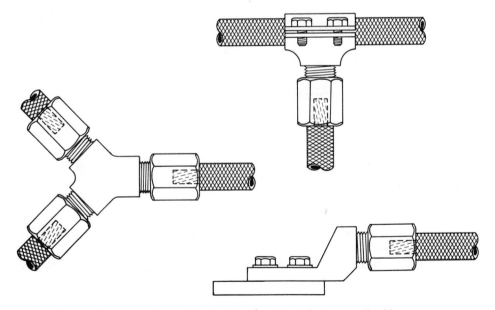

Figure 16-5 Aluminum solderless lugs for connections to small cables

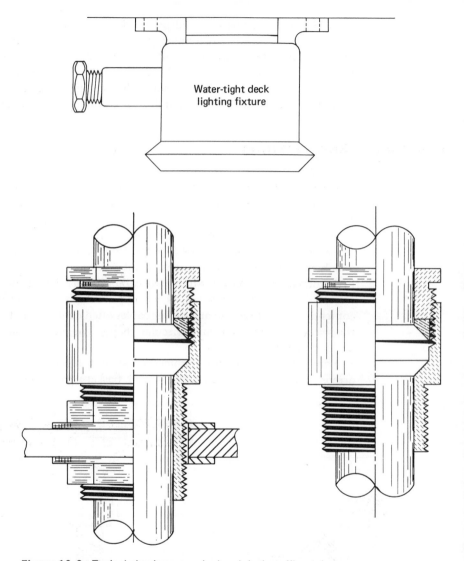

Figure 16-6 Typical aluminum terminal and deck stuffing tubes

threads are thin, care must be taken during installation to avoid dragging the
armored cable over sharp edges and cutting the metallic fabric. When installed through decks, armored cable is enclosed in conduit for additional
protection. Armored cable is available in various types that resist moisture,
oil, heat, or fire. Basket-weave cable is widely used in marine work because of
its comparatively light weight, ruggedness, ease of installation, flexibility, and
serviceability. Up to 1,000,000 feet of cable may be installed in a large ship.

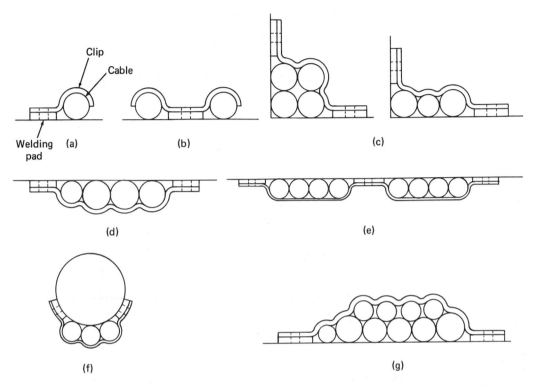

Figure 16-7 Basic types of cable hangers (a) single clip (b) double clip (c) corner hangers (d) 4-way hanger (e) two-span hanger (f) stanchion hanger (g) banked hanger

Aluminum fittings are generally utilized because of their light weight. Figure 16-5 shows typical solderless lugs used for making connections to small cables. Fixtures and other devices are sealed tightly by means of suitable fittings where cables enter. For example, Figure 16-6 depicts typical terminal and deck stuffing tubes or packing tubes. Various types of hangers are utilized for supporting cable runs. As an illustration, Figure 16-7 (a–g) shows typical hangers mounted on pads. Single or double clips mounted on pads may be used for one or two wires less than 1 inch in diameter. Corner hangers mounted on pads can accommodate several wires. A four-way hanger is shaped to support four cables. A pair of four-way hangers to support eight cables is called a two-span hanger. Stanchion hangers are formed to provide cable support on curved surfaces. A banked hanger accommodates two layers of cables.

Most hangers are installed by welding, although a hanger may also be secured by machine screws. When cables are installed in locations exposed to the weather, or in fire or engine rooms, the hangers are mounted on brackets. It is sometimes necessary to install brackets to provide clearance between

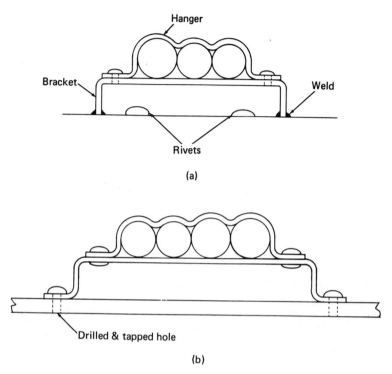

Figure 16-8 Basic bracket installations (a) welded to supporting surface (b) drill-and-tap (machine screw) method

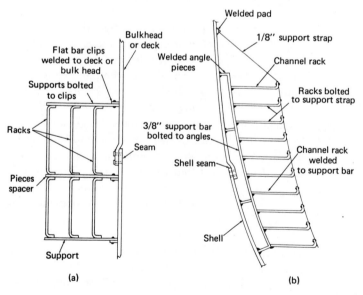

Figure 16-9 Basic types of cable racks (a) deck-type or bulkhead rack (b) shell rack

the hanger and some mechanical obstruction, such as rivets. Figure 16-8 depicts basic bracket arrangements. The hanger is generally secured to the bracket with machine screws. In turn, the bracket is secured to the mounting surface either by welding or by machine screws. Brackets and hangers are generally made up as required by the shipyard machine shop. They are fabricated as light in weight as possible to minimize the total weight of the wiring installation.

When too many cables must be run in the same direction to be supported by a conventional hanger, a cable rack is utilized. Most cable racks can be described as deck-type or shell-type racks. Examples are shown in Figure 16-9. Deck-type racks are installed in passageways and similar locations. Shell-type racks are installed in engine rooms and similar compartments. Racks may be made of either aluminum or steel. They are generally installed at 15-inch intervals, as indicated in Figure 16-10. Note that the curvature of the shell plates changes from one bulkhead to the next, and the curvature of shell-type cable racks must also be changed to match the shell. Racks are welded to the shell. Note in Figure 16-9 that deck-type racks are generally bolted to clips that in turn are welded to a bulkhead or deck surface. Bolting permits the rack to be easily replaced by another size of rack, if this should become necessary.

Figure 16-11 shows the fixtures and lighting circuit for a typical stateroom. When armored cable is utilized, the armor should not protrude inside the fixture where its sharp edges could cut into insulation. When insulation is removed from wires for making connections, none of the strands should be

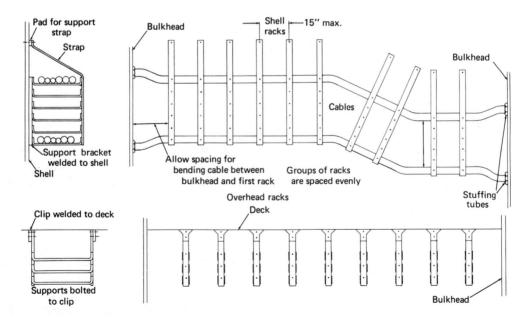

Figure 16-10 Installation of side or shell racks

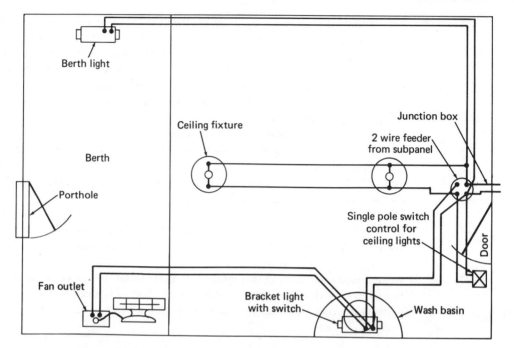

Figure 16-11   Fixtures and lighting circuit for a typical stateroom

cut or nicked. A small lug may be soldered at the end of the wire for making a secure connection. If lugs are not used, the end of a wire may be formed into an "eye" and thoroughly tinned with solder. Any exposed area of the conductor should then be taped and coated with insulating paint. When the "eye" or lug is connected into the fixture, the terminal screw should be turned down tightly, but excessive force that might strip the threads must be avoided.

16–5 MISCELLANEOUS REQUIREMENTS

When parallel conduits are installed, the unions, couplings, etc. are placed in line, provided accessibility is not hampered. In case the conduits are run comparatively close to one another, it becomes advisable or necessary to stagger the unions, as depicted in Figure 16-12. This provides maximum accessibility, and if the unions are placed in two even lines, the finished job will appear quite workmanlike. All conduits must be sufficiently large to permit pulling the cables through without exerting excessive force. Gradual bends in the conduit also contribute to ease of cable pulling. Graphite or equivalent lubricant may be introduced into conduit if necessary to minimize pulling friction.

Cables installed through decks must be protected from mechanical damage above deck. Therefore, kick pipes are installed, as depicted in Figure 16-13. Kick pipes are approximately 15 inches long. It is essential to make kick

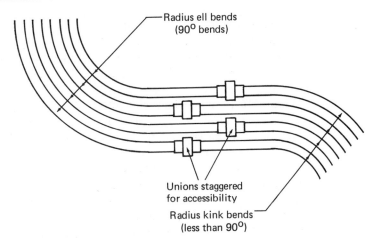

Figure 16-12 Staggering of unions when installing closely-spaced conduits

pipes watertight by means of deck packing tubes, as noted previously. Several cables may be run through each kickpipe. Bends in kickpipes should be neatly made so that a workmanlike appearance is provided; uniformity and symmetry are the basic requirements. Hangers mounted on welding pads

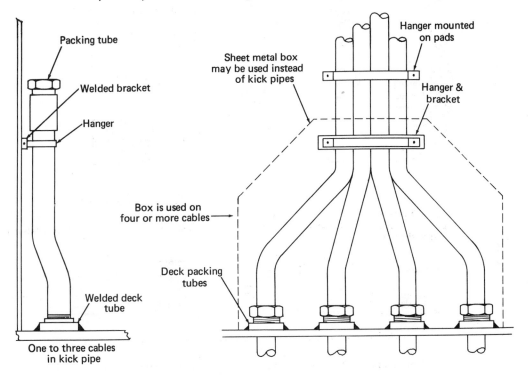

Figure 16-13 Installation of kick pipes

or brackets are installed at the tops of the kickpipes to provide rigid mechanical support. If a large number of cables run through a deck, they are individually sealed with deck packing tubes, but a sheet-metal watertight box may be installed instead of numerous kick pipes. The height of a box is generally 15 inches.

Figure 16-14 shows how 480-volt cables are prepared for connections. The first step is to cut each conductor about 3 inches longer than is absolutely necessary. The outer insulation is then carefully removed, as shown in the diagram, with a sharp knife. Any jute or other interior filler is also removed. Then the conductors are separated for taping. Each conductor is wrapped first with oiled linen tape and then with friction tape or approved plastic tape. These layers of tape are brought up to a depth such that the taped conductors together have the same diameter as the outer insulation that was removed. The cable is then placed carefully in position and the conductors

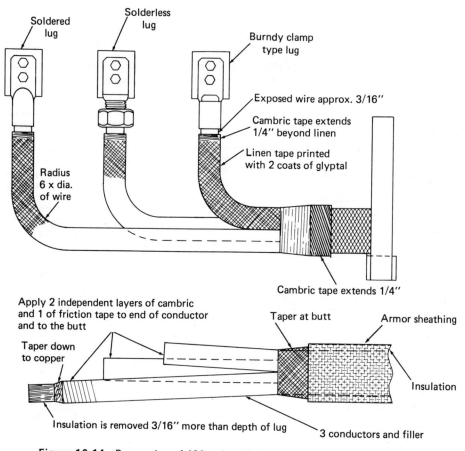

Figure 16-14 Preparation of 480-volt cable for connections

are cut to the exact lengths required for each connection. Each conductor is then stripped of its insulation for a length 1/4 inch greater than the depth of the lug. All strands must be well cleaned without mechanical damage.

Next, the lugs are removed from the terminal block and cleaned preparatory to soldering. The soldering operation requires a gas torch, soldering pot, ladle, and gasoline torch. The strands of a conductor are first tinned over a pot by pouring molten solder over the fluxed wires. A minimum amount of flux should be utilized. Next, the inside surfaces of the lug are tinned in a similar manner. Then, the lug is filled with molten solder. The tinned end of the conductor is inserted into the lug and additional solder is poured into the lug until it is completely filled. It is good practice to let lugs cool slowly. If the cooling solder contracts so that the lug is not completely filled, the gasoline torch is used to melt additional wire solder into the lug.

Note that the oiled linen tape is wrapped to a depth on the conductor equal to the diameter of the lug. In addition, two or three more layers are wrapped about two-thirds of the way down the exposed body of the lug. Then the top layer of friction or plastic tape is started about one-third of the way down the lug so that the oiled linen tape remains exposed on the lug past the outer layer of tape. Finally, the tape is coated with an insulating compound such as Glyptal. This coating provides moisture-proofing. It is essential to cut and prepare cables carefully because any error that may be made requires removal of the defective cable and pulling a new cable through the run of conduit. This reworking is both costly and time-consuming.

QUESTIONS

1. How many conductors are in the cable run from the feeder distribution panel to the distribution panel on the ship?
2. How are motor generators connected when used in welding units?
3. What is the current range of a 70-volt welder?
4. What is the supply that shipyards usually employ?
5. Is the cable on a 180-Hz cable heavier or lighter than that of a 60-Hz cable?
6. What type of insulation is widely used in marine installations?
7. What type of cable is used to provide protection against mechanical damage?
8. What type of fittings are generally used?
9. What is the name of a pair of four-way hangers that are used to support light cables?
10. What type of rack is installed in engine rooms and similar compartments?
11. What is the name of the pipes that are used above deck to protect cables that are installed through decks?
12. What is the name of a popular insulating compound that is used for moisture-proofing?

17 AVIATION WIRING

17–1 GENERAL CONSIDERATIONS

Both AC and DC power are utilized in aircraft electrical systems. Generators and storage batteries are used as DC sources, and inverters are employed as AC sources at frequencies from 60 to 800 Hz. Aircraft storage batteries consist of 3, 6, or 12 cells for operation in 6–8-, 12–16-, and 24–28-volt systems. Inverters typically provide 115-volt 400-Hz three-phase output. There has also been a trend to 208/120 volt, four-wire, three-phase AC systems. With reference to Figure 17-1, an aircraft wiring system comprises bus and feeder circuits. The similarity to industrial installations is notable in large aircraft wiring systems. Ground return circuits are utilized in all-metal aircraft. Negative DC-grounding is standardized. In a two-wire DC system, the negative terminal of the battery is connected to available metal structures. One phase of an AC system, or the neutral wire of a four-wire system, is similarly grounded.

Power and lighting circuits in aircraft are installed with stranded copper cable in gauges from 4/0 to 22. Ignition cable ranges from 5 to 9 millimeters in diameter and is made from stranded copper or steel conductors. Aircraft cables are designed for high mechanical strength, flexibility, and ability to withstand heat, moisture, and fungus. Polyvinyl chloride plastic insulation is typical, with an outer cover of cotton, rayon, glass, or asbestos. This covering is impregnated with lacquer or varnish. Aluminum conductors are employed in larger sizes of cables. Individual wires are terminated with terminal lugs. Junction panels are provided in the aircraft structure for bolting the lug of one conductor to the lug of another conductor. A junction panel consists of a strip of insulating substance with a number of terminal posts. It may be enclosed in a metal box or concealed behind a heavy canvas cover.

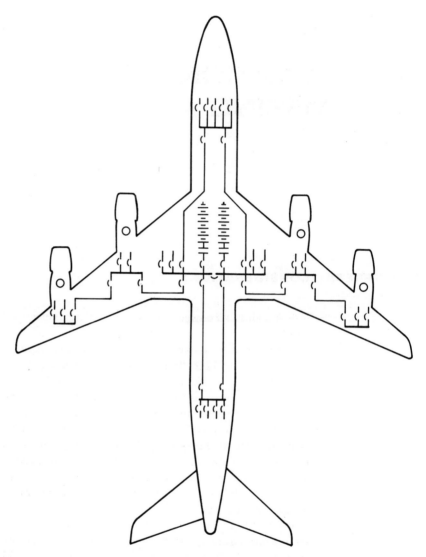

Figure 17-1 Typical bus and feeder arrangements in aircraft

17–2 TERMINALS AND CONNECTORS

Both soldered and solderless terminals are used to terminate conductors, although solderless terminals are preferred. A solderless terminal is installed as depicted in Figure 17-2. Connectors are terminal devices that make

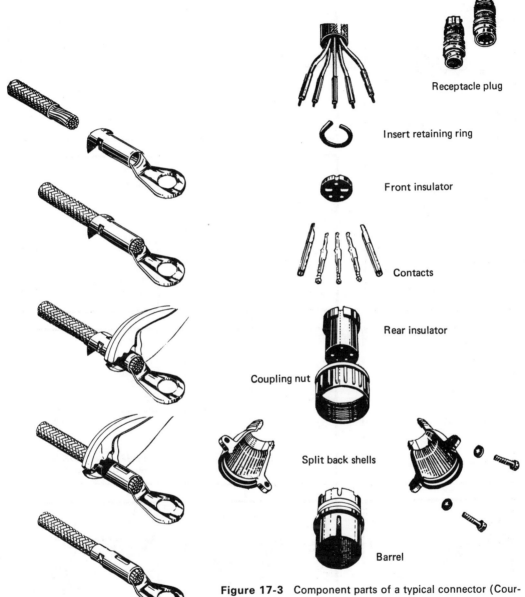

Receptacle plug

Insert retaining ring

Front insulator

Contacts

Rear insulator

Coupling nut

Split back shells

Barrel

Figure 17-3 Component parts of a typical connector (Courtesy U.S. Armed Forces)

Figure 17-2 Installation of a solderless terminal (Courtesy U.S. Armed Forces)

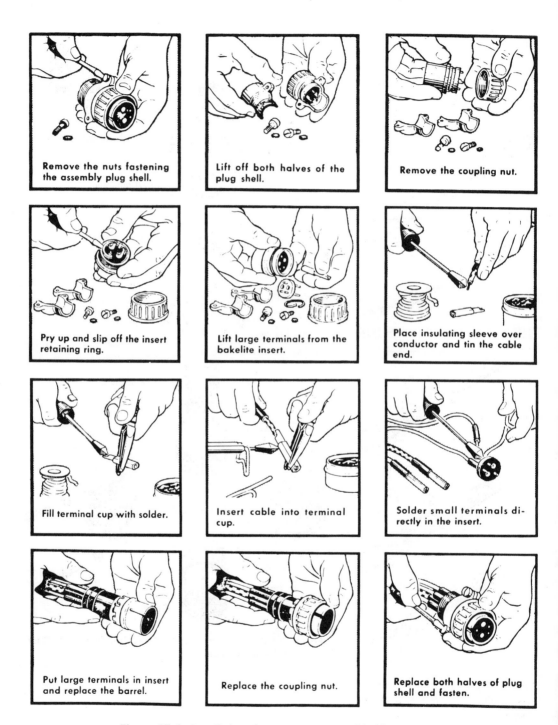

Figure 17-4 Installation of a connector on a cable (Courtesy U.S. Armed Forces)

cables or sets of wires easy to connect or disconnect. Figure 17-3 shows the component parts of a typical connector. It consists of a fixed portion called the receptacle and a movable portion called the plug. Plugs are available in either straight or 90° angle types. Receptacles are of the wall mounting, box mounting, or integral mounting types. Figure 17-4 shows the procedure for installing a connector on a cable.

17–3 BONDING

Bonds are connections that are used to tie together electrically any metal parts of an aircraft that are not already integral parts of the metal structure. As an illustration, bonds are used to connect an engine to a mount, an aileron to a wing, a radio to the main structure, a cowling flap to a support, and conduit (if used) to the aircraft structure. A bond usually consists of a flexible metal strap provided with a terminal at each end. It may be made of either aluminum or tinned copper wire. Note that aluminum should not be connected in contact with a dissimilar metal. That is, a contact of brass, bronze, copper, or steel with aluminum will result in corrosion, particularly in the presence of moisture. It is essential for a bond to be mechanically and electrically reliable, so that a good electrical contact is maintained at all times.

17–4 WIRING DIAGRAMS

A wiring diagram of the electrical system is provided with an airplane. It should be kept in the data compartment of the plane. The diagram may be a blueprint or, in the case of a large aircraft, it will consist of a number of prints bound in book form. Symbols used in aircraft wiring diagrams are shown in Figure 17-5. Part of a wiring diagram is illustrated in Figure 17-6. Dashed lines in a diagram may show in a general way the main sections of the airplane and wiring system. They may also denote junction boxes, other enclosures, and sometimes mechanical linkages. Diagrams in book form provide separate schematics for each circuit. That is, the lighting, power, starting, and other circuits are each placed on a single page or on adjacent pages.

In addition to the symbols shown in Figure 17-5, various other symbols that designate items of electrical equipment appear on aircraft schematics. An example is given in Figure 17-7. A reference number is assigned to each separate item and is printed near the symbol that represents the item. A table of equipment, on which each item is listed numerically, may be pro-

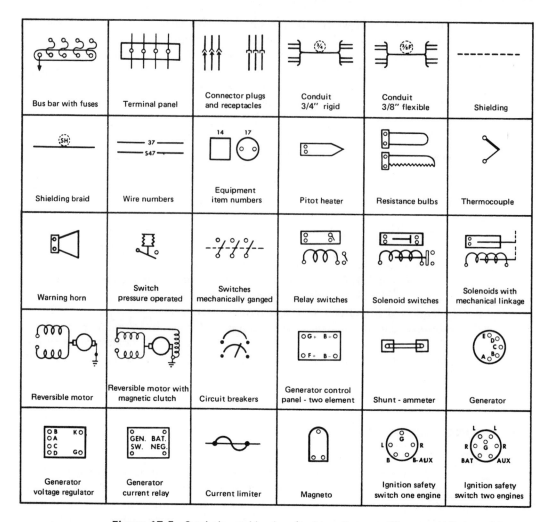

Figure 17-5 Symbols used in aircraft wiring diagrams (Courtesy U.S. Armed Forces)

vided on the blueprint. The information given for each item includes its serial number, name or description, quantity required, and so on, as shown in the list of equipment in Table 17-1. Cables which connect various units are indicated by solid vertical or horizontal lines. For identification, the serial number on the cable in the diagram is the same as that on the cable that is installed. A wire table, in which the cables are listed numerically, is also provided on the blueprint. The information for each cable may include length, size, and type of terminal lugs.

Conduit is generally represented on aircraft wiring diagrams by a single solid line. At each end of the line a bracket is drawn to include all of the separate cables entering or leaving the conduit. All cables entering a conduit

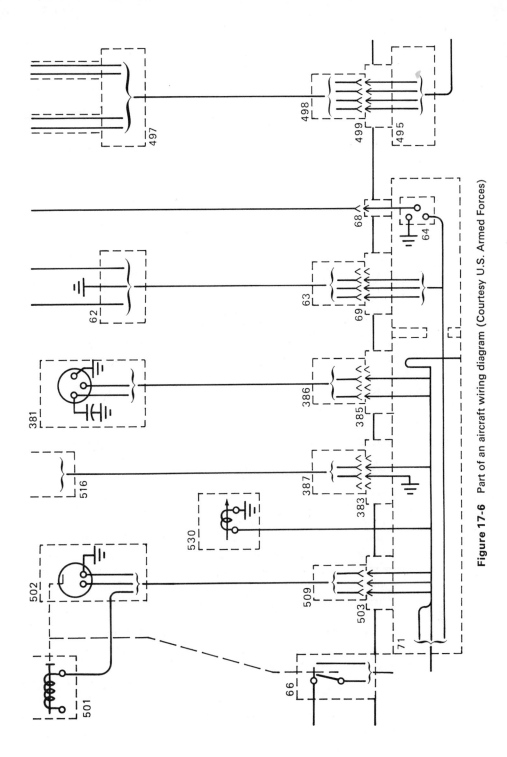

Figure 17-6 Part of an aircraft wiring diagram (Courtesy U.S. Armed Forces)

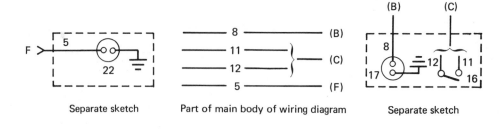

Separate sketch   Part of main body of wiring diagram   Separate sketch

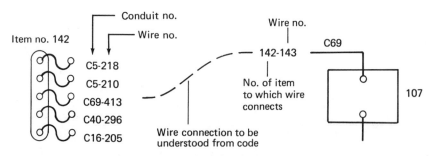

Figure 17-7 Examples of wiring diagram code conventions (Courtesy U.S. Armed Forces)

at one point must leave it at another point, but not necessarily in the same order of entry. Conduit and cable lines should not be confused when reading a schematic. Note that a conduit line may or may not have an equipment number, but it never has a cable number. Occasionally, a unit (or group of units) appears in a separate sketch. Note in Figure 17-7 the illustration of the code conventions method of showing the relationship between the sketch and the main part of the wiring diagram.

Some wiring diagrams employ a complete "code" in order to avoid many crossings of wires. When this is done, only a small length of the cable from its terminals is shown, instead of drawing the entire run, and at each terminal of an item of equipment various code numbers are given. One number represents the wire connected to a terminal, and another (adjacent to the cable number) designates the item of equipment to which the cable connects. When tracing a cable on this type of diagram, locate the item of equipment first and then the terminal to which the cable that you are tracing is connected. Individual circuit diagrams, as shown in Figure 17-8, are often provided in addition to a complete wiring diagram. This type of diagram is easier to use when the electrician is concerned with only one particular circuit.

An example of a detailed diagram with a simplified diagram is shown in Figure 17-9. A simplified diagram facilitates understanding the system

Table 17-1 Arrangement of an aircraft equipment and wire table (Courtesy U.S. Armed Forces)

TABLE OF EQUIPMENT

No.	Description	Req.	Type	Part no.	Inst. dwg. no.
61	Magneto — Bendix-Scintilla	8	SF9L-1	10-5377	15-5900
62	Shield — Engine Pull	4	BAC	3-10569-12	6-8107-1
63	Plug — Ignition Pull	4	BREEZE	35A-4273	15-5284
64	Coil — Booster	4	A-1	2520	
65	Shield Outbrd. Ignition	2	BAC	3-1256	6-7692
66	Shield Booster Switch (Outbd.)	2	BAC	21-9627	59-2917 69-2917

381	Generator — Engine Nos. 2-3 & 4	3	E-7	F-64463-2	15-5900
382	Plug	4	BREEZE	35A4273	15-5284
383	Socket	4	BREEZE	35A4269	
384	Capacitor	7	B-1	Spec. 50131	
385	Socket — Generator	3	BREEZE	35B5168	3-11255

491	Thermocouple — Unit	4	W. A. C.	62508	62102
494	Resistance Bulb — Carb. Air Temp.	4	F-2	Spec. 27823	15-5900
495	Shield — Outbrd. Thermocouple	2	BAC	3-10569-6 3-10569-12	3-11377
496	Shield — Inbrd.	2	BAC	"	3-11969
497	Shield — Engine Pull	4	BAC	"	6-8107
498	Plug	4	BREEZE	35A2533	15-5284
499	Socket	4	BREEZE	35A2539	3-11969

WIRE TABLE — IGNITION

Wire no.	Size	Length
I 1	16	264
I 2	16	48
I 21	18	312
I 22	18	48
I 23	HT	6

WIRE TABLE — D. C. POWER

Wire no.	Size	Length
P 2	18	60
P 62	18	372

WIRE TABLE — STARTER CONTROL

Wire no.	Size	Length
S 21	10	372
S 23	16	288
S 24	2	96
S 31	2	120

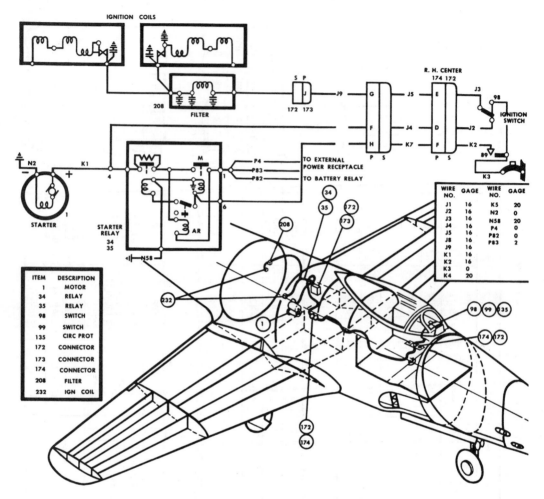

Figure 17-8 Example of an individual circuit diagram for the starter and ignition circuit (Courtesy U.S. Armed Forces)

operation, whereas a detailed diagram depicts each circuit connection that is required. It is helpful to use a simplified diagram when a trouble symptom occurs in a wiring system and the symptom must be analyzed to close in on the defective connection, wire, or component. Another useful type of diagram in this situation is a layout of a single-engine aircraft electrical system, as given in Figure 17-10. Simplified diagrams and layout diagrams may or may not be provided by the manufacturer of an aircraft. However, the electrician can make up such diagrams for himself by studying the wiring blueprint for the aircraft.

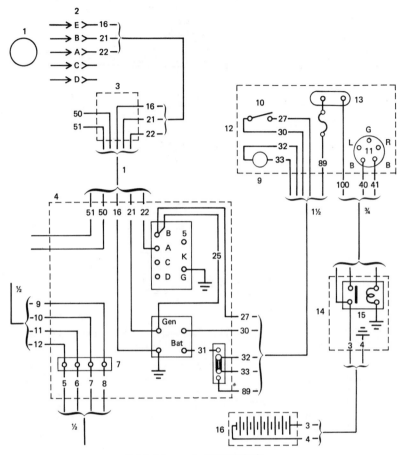

Detailed diagram of a generator circuit

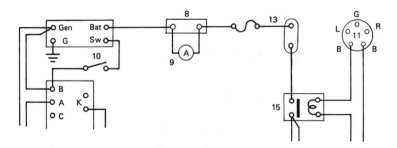

Simplified diagram of a generator system

Figure 17-9 Example of a simplified diagram with a detailed diagram (Courtesy U.S. Armed Forces)

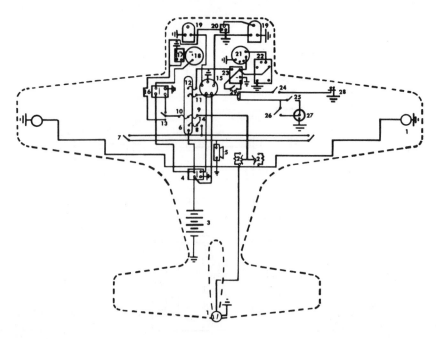

1. Position light	11. Fuse, fuel-pressure warning	20. Booster coil
2. Switch, position light	and pitot heater	21. Generator
3. Battery	12. Fuse, generator	22. Voltage regulator
4. Relay battery	13. Switch, starter	23. Relay switch
5. Warning horn	14. Switch, throttle	24. Switch, pitot heater
6. Horn	15. Switch, ignition	25. Switch, fuel-pressure warning
7. Switch, landing-gear warning	16. Relay, starter	26. Switch, fuel-pressure test
8. Fuse, landing-gear warning	17. Solenoid, starter meshing	27. Fuel-pressure warning light
9. Fuse, position light	18. Starter	28. Pitot heater
10. Fuse, starter	19. Magneto	29. Switch, generator

Figure 17-10 A diagrammatic layout of basic electrical items (Courtesy U.S. Armed Forces)

17–5 AIRPLANE LIGHTING SYSTEMS

An aircraft lighting system has exterior lights for illumination at night in operations such as landing. Interior lighting provides illumination for instruments, equipment, cockpits, and passengers. Indicator lights inside the plane reveal the status of equipment, such as the position of the landing gear. The principal types of exterior lights are the navigation or position lights, recognition lights, landing lights, and passing lights. A typical navigation light circuit is shown in Figure 17-11. At least one set of navigation lights is installed on the exterior of the craft. A set of these lights consists

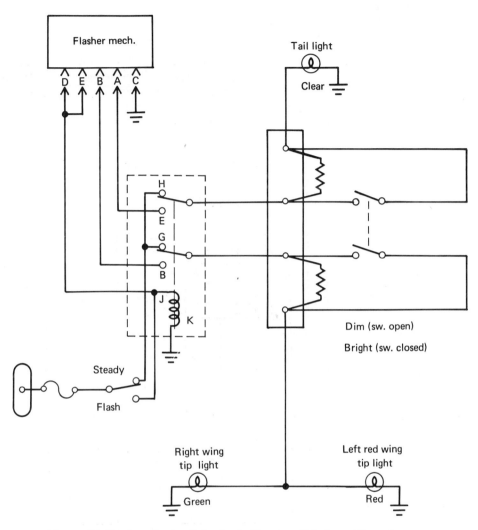

Figure 17-11 A navigation light circuit (Courtesy U.S. Armed Forces)

of one red, one green, and one white unit. The green unit is mounted at the extreme tip of the right wing, and the red unit is mounted in a similar position on the left wing. The white unit is usually located on the vertical stabilizer in a position where it is clearly visible through a wide angle from the rear of the aircraft.

The wing-tip lamp and the tail lamp can be controlled by a DPST switch in the pilot's compartment. By one of two "on" positions, Dim and Bright, the pilot can choose between two brightness levels. On the Dim position, the switch connects a resistor in series with the lamps; on the Bright position, the resistor is shorted out and the lamps operate at full brightness. A switch

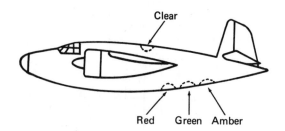

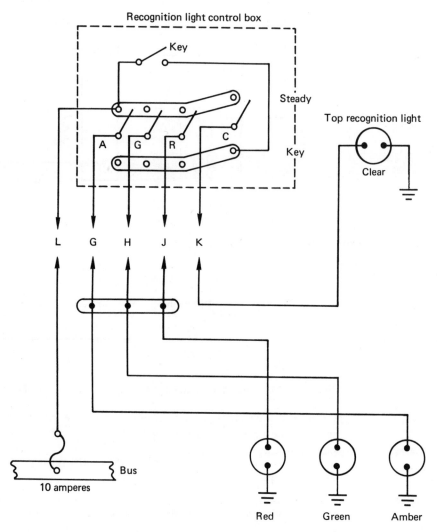

Figure 17-12 Arrangement of recognition lights (Courtesy U.S. Armed Forces)

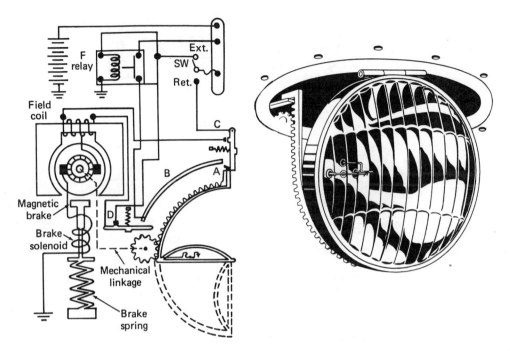

Figure 17-13 Landing-light circuit and control mechanism (Courtesy U.S. Armed Forces)

in the pilot's compartment provides for momentary, steady, flashing, or off operation of the navigation lights. For flashing operation, a flasher mechanism is installed in the navigation light circuit. Although the number of recognition lights may vary, there are usually three or four. They are usually located on the underside of the fuselage or the right wing. A fourth light, if used, is located on the top of the fuselage. (See Figure 17-12.)

Landing lights are installed to illuminate runways during night landings. They are located either midway in the leading edge of each wing or streamlined into the surface of the aircraft. Each light may be controlled by a relay or it may be connected directly into the electrical circuit. Since icing of the lamp lenses reduces illumination, some installations utilize retractable-type landing lamps. When the lamps are not in use, a landing-light motor retracts them into receptacles in the wing where the lenses are not exposed to the weather. Figure 17-13 depicts a landing-light circuit and control mechanism. A motor operates a pinion-and-rack gear to raise and lower the light. A magnetic brake is employed to stop the motor promptly when the cycle of operation is completed.

In addition to the usual wing-tip navigation lights, a passing light is sometimes included in the lighting system as a precaution against collision

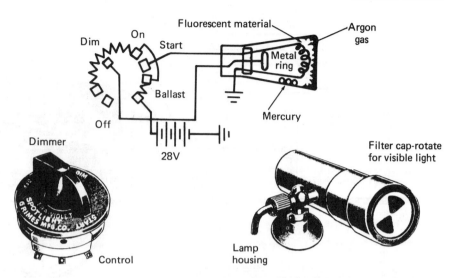

Figure 17-14   Circuit for instrument lamp assembly (Courtesy U.S. Armed Forces)

when meeting other aircraft. This lamp is installed in the leading edge of the left wing and has a red lens. It is controlled by a toggle switch in the pilot's compartment.

Various types of lights are employed in the interior of an aircraft. Lamps set in the instrument panel, which is covered by a reflector panel, may provide a means of indirect lighting. The light from these lamps is distributed over the entire panel. The reflector has openings in it for observing the various instruments. Another form of instrument lighting uses fluorescent lamps to eliminate glare. The lamp assembly has a special lens which will pass only ultraviolet light, with a screen to regulate the amount of light. Another screen passes the visible light. Instruments used with this type of lighting have dial figures painted with a material that is sensitive to ultraviolet light. In turn, the figures glow when struck by the invisible ultraviolet rays.

The fluorescent lamp assembly shown in Figure 17-14 operates directly on the 28-volt DC line. It consists of a small fixture containing a special 4-watt, 28-volt fluorescent lamp equipped with a light filter that transmits only the invisible ultraviolet rays. To obtain visible light, the filter is opened. A combined starting switch and intensity rheostat mounted on the instrument panel controls the lamp. When the switch is thrown to its Start position, a circuit is completed through the battery, switch, and bulb filament, heating the gas and vaporizing the mercury in the bulb. The gas and mercury vapor serve as a conductor and, when the switch is placed in its On position, complete a circuit to ground through the battery, switch, metal ring, and filament. Like all electrical lamps, the fluorescent unit must eventually be replaced.

QUESTIONS

1. What is the frequency range of aircraft inverters?

2. What polarity of DC-grounding is standard?

3. What is the name of a typical insulation used in aviation wiring?

4. What type of terminals are preferred in aircraft wiring?

5. What does a bond usually consist of in electrical wiring?

6. How is conduit generally represented on aircraft wiring diagrams?

7. What are the colors of a set of navigation lights?

8. Where are recognition lights usually located?

18 TELEVISION WIRING

18–1 GENERAL CONSIDERATIONS

Community antenna television (CATV) and master antenna television (MATV) installations provide a number of TV receivers with signals from a single antenna system. In fringe areas it is advantageous to install an antenna system in a carefully selected location, and to employ a cable network from the antenna site to the receivers that are served. An MATV system generally serves a multiple-occupancy building such as an apartment house or hotel. A CATV system usually serves an entire town or city. There is a trend to the installation of CATV systems in cities that are plagued with ghosts, because the antenna system can be located at a ghost-free site. In any case, the basic purpose of a CATV or MATV system is to provide snow-free and ghost-free high-quality TV signals to subscribers in difficult reception areas.

Signals supplied by MATV and CATV systems generally comprise VHF, UHF, and FM transmissions. Since a CATV system usually involves long runs of cable, efficiency is improved by heterodyning the ultra-high frequency (UHF) signals to lower frequencies in the very-high frequency (VHF) band. The heterodyne units that are used are called translators. Although an antenna is installed at a strategic site, it cannot intercept sufficient signal to supply a large number of receivers. In turn, amplifiers are employed to increase the signal power as required. Amplifiers are also installed at intervals along extended cable runs to compensate for line losses. When two TV receivers are connected to a common signal source, they will interfere with each other unless some means of isolation is utilized. The simplest isolation method is to insert a loss pad between each receiver and the line. According-

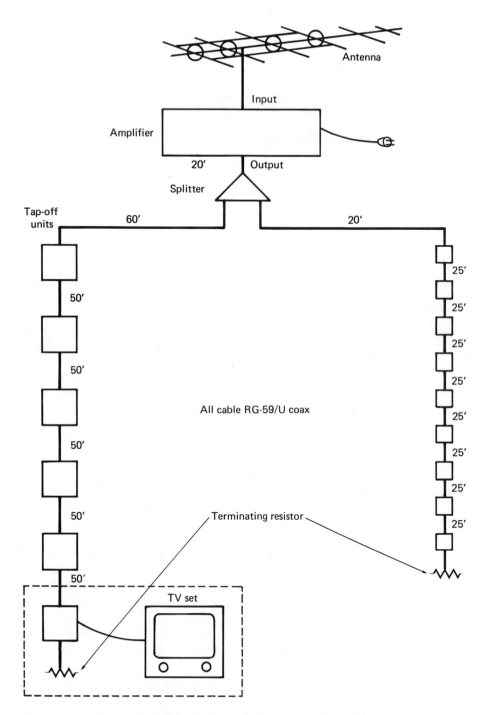

Figure 18-1 Block diagram of a basic MATV distribution system

ly, additional amplification is required to overcome this loss. Figure 18-1 shows the arrangement of a basic MATV system.

A distribution system consists of an antenna (or antennas) to intercept the signals, amplifiers to compensate for system losses, transmission lines (coaxial cables) to conduct the signal currents from source to loads, splitters to branch the lines, and tap-off units to provide isolation between receivers. Most installations employ 75-ohm coaxial cable in the system network.

Table 18-1   TV channel allocations

P / S Freq.	Ch. No.	Freq. Limits		P / S Freq.	Ch. No.	Freq. Limits		P / S Freq.	Ch. No.	Freq. Limits
		54		P 543.25 / S 547.75	26	542 / 548		P 717.25 / S 721.75	55	716 / 722
P 55.25 / S 59.75	2	60		P 549.25 / S 553.75	27	548 / 554		P 723.25 / S 727.75	56	722 / 728
P 61.25 / S 65.75	3	66		P 555.25 / S 559.75	28	554 / 560		P 729.25 / S 733.75	57	728 / 734
P 67.25 / S 71.75	4	72		P 561.25 / S 565.75	29	560 / 566		P 735.25 / S 739.75	58	734 / 740
		76		P 567.25 / S 571.75	30	566 / 572		P 741.25 / S 745.75	59	740 / 746
P 77.25 / S 81.75	5	82		P 573.25 / S 577.75	31	572 / 578		P 747.25 / S 751.75	60	746 / 752
P 83.25 / S 87.75	6	88		P 579.25 / S 583.75	32	578 / 584		P 753.25 / S 757.75	61	752 / 758
		174		P 585.25 / S 589.75	33	584 / 590		P 759.25 / S 763.75	62	758 / 764
P 175.25 / S 179.75	7	180		P 591.25 / S 595.75	34	590 / 596		P 765.25 / S 769.75	63	764 / 770
P 181.25 / S 185.75	8	186		P 597.25 / S 601.75	35	596 / 602		P 771.25 / S 775.75	64	770 / 776
P 187.25 / S 191.75	9	192		P 603.25 / S 607.75	36	602 / 608		P 777.25 / S 781.75	65	776 / 782
P 193.25 / S 197.75	10	198		P 609.25 / S 613.75	37	608 / 614		P 783.25 / S 787.75	66	782 / 788
P 199.25 / S 203.75	11	204		P 615.25 / S 619.75	38	614 / 620		P 789.25 / S 793.75	67	788 / 794
P 205.25 / S 209.75	12	210		P 621.25 / S 625.75	39	620 / 626		P 795.25 / S 799.75	68	794 / 800
P 211.25 / S 215.75	13	216		P 627.25 / S 631.75	40	626 / 632		P 801.25 / S 805.75	69	800 / 806
		470		P 633.25 / S 637.75	41	632 / 638		P 807.25 / S 811.75	70	806 / 812
P 471.25 / S 475.75	14	476		P 639.25 / S 643.75	42	638 / 644		P 813.25 / S 817.75	71	812 / 818
P 477.25 / S 481.75	15	482		P 645.25 / S 649.75	43	644 / 650		P 819.25 / S 823.75	72	818 / 824
P 483.25 / S 487.75	16	488		P 651.25 / S 655.75	44	650 / 656		P 825.25 / S 829.75	73	824 / 830
P 489.25 / S 493.75	17	494		P 657.25 / S 661.75	45	656 / 662		P 831.25 / S 835.75	74	830 / 836
P 495.25 / S 499.75	18	500		P 663.25 / S 667.75	46	662 / 668		P 837.25 / S 841.75	75	836 / 842
P 501.25 / S 505.75	19	506		P 669.25 / S 673.75	47	668 / 674		P 843.25 / S 847.75	76	842 / 848
P 507.25 / S 511.75	20	512		P 675.25 / S 679.75	48	674 / 680		P 849.25 / S 853.75	77	848 / 854
P 513.25 / S 517.75	21	518		P 681.25 / S 685.75	49	680 / 686		P 855.25 / S 859.75	78	854 / 860
P 519.25 / S 523.75	22	524		P 687.25 / S 691.75	50	686 / 692		P 861.25 / S 865.75	79	860 / 866
P 525.25 / S 529.75	23	530		P 693.25 / S 697.75	51	692 / 698		P 867.25 / S 871.75	80	866 / 872
P 531.25 / S 535.75	24	536		P 699.25 / S 703.75	52	698 / 704		P 873.25 / S 877.75	81	872 / 878
P 537.25 / S 541.75	25	542		P 705.25 / S 709.75	53	704 / 710		P 879.25 / S 883.75	82	878 / 884
				P 711.25 / S 715.75	54	710 / 716		P 885.25 / S 889.75	83	884 / 890

P = Picture Carrier Freq.   S = Sound Carrier Freq.   All frequencies in MHz
VHF Channels 2-13   UHF Channels 14-83

Planning a distribution system starts with consideration of the field strengths (signal strengths) that are available. For example, with reference to Figure 18-1, the intercepted signals might have a strength of 1,000 microvolts each on channels 2, 5, 7, and 13. Field strengths are measured with a field-strength meter, which is a tuned VHF/UHF microvolt meter. Channel allocations are shown in Table 18-1.

A field-strength meter is used, while an antenna is being installed, to determine the orientation that provides maximum signal strength. With reference to Figure 18-1, cable lengths are specified between the various devices. This is an important consideration and, when cable lengths are determined from inspection of building plans or of a building under construction, the required lengths must be carefully calculated, taking into account any added lengths needed in routing from one point to another. That is, when obstructions occur, a straight-line installation is not possible. Thus, if it happens that twice as long a run is required in practice as for an ideal straight-line run, the cable loss will be doubled and proportionally greater amplification must be provided.

18–2 SYSTEM ANALYSIS

The system depicted in Figure 18-1 is analyzed from the viewpoint of its function—feeding signals to TV receivers. If we select a point at which a TV receiver "looking back" at the antenna through the system encounters the most loss-producing components, we can evaluate the maximum demand on the amplifier. In this example, the point occurs at tap-off unit m. That is, the taps to be checked in this system are either s or m. The splitter at d divides the amplified antenna signal; losses are equal at both legs of the system. Therefore, we need to check only the losses between m and the splitter, and between s and the splitter. Since we need to know the amount of loss, we must choose the components to be installed, and consider their associated losses.

Table 18-2 shows a typical isolation and feed-through loss chart. Note that three models are listed: white, red, and yellow. The isolation and feed-through losses for each are listed in decibels (dB). Since amplifier gain is specified in dB units, this is the preferred way of evaluating losses. Note in passing that dB are related to voltage, current, and power ratios, as tabulated in Table 18-3. Isolation losses are those seen between the line and the TV receiver—isolation losses are not seen along the line. Feed-through losses are caused by insertion of the tap-off unit itself, and are added to the line losses. That is, isolation losses do not add to one another, but feed-through losses do. Signals undergo loss in flowing through cables from the amplifier and are weak near the ends of the system, although they are strong near the amplifier. In turn, high isolation values (lower feed-through losses)

Table 18-2 Typical isolation and feed-through loss chart (Courtesy Winegard Antenna Systems)

ISOLATION AND FEED-THRU LOSS CHART					
A = Isolation B = Feed-Thru Loss (values in dB)					
ULTRA-TAP, MODELS UT-22 AND UT-33					
FREQUENCY	CHANNEL NUMBER		WHITE (W)	RED (R)	YELLOW (Y)
54 MHz	2	A	35	27	22.5
		B	0.2	0.4	0.2
88 MHz	6	A	34	25	19
		B	0.2	0.4	0.3
174 MHz	7	A	31	21	13
		B	0.25	0.5	0.4
216 MHz	13	A	30	20	11.5
		B	0.25	0.5	1.0
ULTRA-TAP, MODELS UT-82-CAC AND UT-82-59					
54 MHz	2	A	34	26.5	22
		B	0.5	1.0	1.25
216 MHz	13	A	29.5	25.5	20.5
		B	0.5	1.0	1.25
470 MHz	14	A	29.5	24	17
		B	0.5	1.0	1.25
800 MHz	69	A	29.5	21	14
		B	0.5	1.0	1.25
890 MHz	83	A	29.5	21	14
		B	0.75	1.0	1.5

are used near the amplifier, whereas lower isolation losses (higher feed-through losses) are utilized at the ends of the system.

Losses of signal occur in the antenna, VHF distribution amplifier, splitter (four-outlet amplifier), and in-line tap-off unit. Losses differ in the same unit for different channels. Isolation values are higher on the low TV channels. The reason for this is that cable loss increases at higher frequencies. High channels need to have less isolation between the line and receiver than do low channels. In turn, an MATV system is planned for operation on channel 13, which is the worst-case VHF condition. Table 18-4 lists the loss in dB per 100 feet for the commonly used types of coaxial cable. (See Figure 18-2.)

Table 18-3 Corresponding dB, voltage, current, and power values (Courtesy Triplett, Inc.)

Decibel Table

Power ratio	Voltage and current ratio	Decibels	Power ratio	Voltage and current ratio	Decibels
1.0000	1.0000	0			
1.0233	1.0116	0.1	19.953	4.4668	13.0
1.0471	1.0233	0.2	25.119	5.0119	14.0
1.0715	1.0315	0.3	31.623	5.6234	15.0
1.0965	1.0471	0.4	39.811	6.3096	16.0
1.1220	1.0593	0.5	50.119	7.0795	17.0
1.1482	1.0715	0.6	63.096	7.9433	18.0
1.1749	1.0839	0.7	70.433	8.9125	19.0
1.2023	1.0965	0.8	100.00	10.0000	20.0
1.2303	1.1092	0.9	158.49	12.589	22.0
1.2589	1.1220	1.0	251.19	15.849	24.0
1.3183	1.1482	1.2	398.11	19.953	26.0
1.3804	1.1749	1.4	630.96	25.119	28.0
1.4454	1.2023	1.6	1000.0	31.623	30.0
1.5136	1.2303	1.8	1584.9	39.811	32.0
1.5849	1.2589	2.0	2511.9	50.119	34.0
1.6595	1.2882	2.2	3981.1	63.096	36.0
1.7328	1.3183	2.4	6309.6	79.433	38.0
1.8198	1.3490	2.6	10^1	100.000	40.0
1.9055	1.3804	2.8	$10^1 \times 1.585$	125.89	42.0
1.9953	1.4125	3.0	$10^1 \times 2.512$	158.49	44.0
2.2387	1.4962	3.5	$10^1 \times 3.981$	199.53	46.0
2.5119	1.5849	4.0	$10^1 \times 6.31$	251.19	48.0
2.8184	1.6788	4.5	10^5	316.23	50.0
3.1623	1.7783	5.0	$10^5 \times 1.585$	398.11	52.0
3.5480	1.8836	5.5	$10^5 \times 2.512$	501.19	54.0
3.9811	1.9953	6.0	$10^5 \times 3.981$	630.96	56.0
5.0119	2.2387	7.0	$10^5 \times 6.31$	794.33	58.0
6.3096	2.5119	8.0	10^6	1,000.00	60.0
7.9433	2.8184	9.0	10^7	3,162.3	70.0
10.0000	3.1623	10.0	10^8	10,000.0	80.0
12.589	3.5480	11.0	10^9	31,623.0	90.0
15.849	3.9811	12.0	10^{10}	100,000.0	100.0

Thus, RG-59/U cable has 5.9 dB loss per 100 feet on channel 13. In practice, this figure is rounded off to 6 dB.

Returning to the tap-off unit, it is easy to decide that if we are considering the receiver at the highest-loss point in the system, we will choose the tap with the lowest isolation loss for channel 13. A suitable unit (Table 18-2) has 11.5 dB isolation loss (yellow unit). It has a 1.0 dB feed-through loss.

Table 18-4 Construction of a coaxial cable showing loss characteristics for different types of coaxial cables (Courtesy Winegard Antenna Systems)

COAXIAL CABLE LOSS (per 100 feet)

CABLE TYPE	VHF CHANNELS (loss—dB)							UHF CHANNELS (loss—dB)		
	2	4	6	FM	7	10	13	14	47	83
RG-59/U*	2.8	3.2	3.6	4.0	5.3	5.6	5.9	9.3	11.8	13.0
RG-6/U*	2.1	2.3	2.6	2.9	4.0	4.0	4.3	6.4	8.4	9.2
RG-11/U*	1.6	1.8	2.0	2.2	2.7	2.7	3.0	4.4	5.8	6.4
RG-59 Foam*	1.8	2.1	2.4	2.7	3.8	3.9	4.0	5.9	7.9	8.6
RG-11 Foam*	1.1	1.2	1.4	1.5	2.1	2.2	2.3	3.4	4.5	4.9
CAC**	1.75	1.9	2.4	2.45	2.8	2.84	3.5	5.5	7.4	8.2

*UHF attenuation figures are nominal calculated values.

UHF attenuation figures are specified **maximum values. Jerrold "CAC" 82-channel coaxial cable is controlled in manufacture to insure the absence of discontinuities from 54 to 890 MHz.

The receiver, therefore, "looks at" an 11.5 dB loss. We will enter this figure on a sheet under the heading "Losses." Two columns are employed: "Losses, Leg s," and "Losses, Leg m," with reference to Figure 18-1. Thus, we write:

Losses, Leg s		*Losses, Leg m*	
Isolation	11.5 dB	Isolation	11.5 dB

Next, looking back through the system, we observe that leg m will have feed-through losses of eight taps (the last one does not show). Leg s will have feed-through losses of five taps. We do not know what color code these taps will have, but we know that they will all be of the yellow type. In turn, an approximation must be made. In practice, most of the taps will be coded red, a few white, and a few yellow. A reasonable average feed-through loss is 0.8 dB. In turn, we proceed with the listing as follows:

Losses, Leg s		*Losses, Leg m*	
Isolation	11.5 dB	Isolation	11.5 dB
Feed-through	4.0 dB	Feed-through	6.4 dB

That is, 5×0.8 dB equals 4.0 dB, and 8×0.8 dB equals 6.4 dB. Next, the cable losses must be added to the list. Adding up the cable lengths, we find that leg s has 310 feet of cable, with a loss of 18.6 dB. Leg m has 220 feet, with a loss of 13.2 dB. These figures are entered in the listings:

Table 18-5 Specifications of typical MATV amplifiers (Courtesy B & K Manufacturing Co.)

	Model 3660	Model 3440
Frequency Range	|← 54 to 108 and 174 to 216 MHz →|	

Minimum Gain
  Lo-Band:
  Hi-Band:
  FM-Band:
	40 dB
	40 dB
	40 dB
		22 dB
		22 dB
		19 dB

Gain Control (minimum)	10 dB	NA

Impedance
  Input:
 	
  Output:
	75 ohms*
	
	75 ohms
		75 or 300 ohms
		(switch selected)
		75 ohms

Noise Figure (maximum)	9 dB	Lo-Band: 5.1 dB Hi-Band: 7 dB
Output Capability per Channel at 0.5% Cross-Modulation**	+50 dBmV, each of any 7 channels +52 dBmV, each of any 3 channels	Lo-Band: +45 dBmV, each of any 3 channels Hi-Band: +43 dBmV, each of any 4 channels
Flatness of Response	1.5 dB P/V	Lo-Band: 1.5 dB P/V Hi-Band: 2 dB P/V
Power Requirement	117 VAC, 60 Hz, 16 watts	117 VAC, 60 Hz, 5 watts
Dimensions	3½″ L x 19″ W x 3¾″ H	7⅜″ L x 5-3/16″ W x 1½″ H
Shipping Weight	6 lbs.	1½ lbs.

*Model 3660 provides switch selection of either dual high-low band inputs or a combined high-low input.

**The units are capable of much higher output operation for the same number of channels; however, 0.5% cross-modulation is well below any objectionable signs of overload and is the best rating to be used in designing a system.

Table 18-5 (cont'd.)

	Model 2880
Frequency Range	54 to 108 MHz and 174 to 216 MHz
Minimum Gain Lo-Band: Hi-Band: FM-Band:	41 dB 44 dB 92 MHz: 41 dB 108 MHz: 20 dB
Impedance Input: Output:	75 ohms 75 ohms
Output Capability Per Channel for 7 Channels at 0.5% Cross-Modulation**	1.0 V (60 dBmV)
Flatness of Response	1.5 dB P/V
Tube Complement	*(1) 6EH7 *(1) 6ES8 *(1) 6DJ8 *frame grid *(2) 6EJ7 types *(3) TJ880 (2) 12BY7A
Power Requirements	117 VAC, 60 Hz 92 Watts
Dimensions	17″ L x 7″ W x 5¼″ H
Shipping Weight	18 lbs.

**Represents capability PER CHANNEL for 3-ch. lo-band and 4-ch. hi-band operation. The unit is capable of much higher output operation for the same number of channels; however, 0.5% cross-modulation is well below any objectionable signs of overload and is the best rating to be used in designing a system.

Table 18-6 Determining signal levels at system tap-off points (Courtesy Winegard Antenna Systems)

Level at amplifier output	50	dBmV	
Loss in cable to splitter (20' of RG-59/U)	—1.2	dB	
Level at splitter (d)	48.8	dBmV	
Loss in splitter	—3.5	dB	
Level at splitter outputs	45.3	dBmV	
Loss to tap-off unit "n" (60' of RG-59/U)	—3.6	dB	
Level at tap-off unit "n"	41.7	dBmV	
(Use white tap, 30 dB isolation)			11.7 dBmV
Feed-thru loss of white tap	—0.25	dB	
Level out of tap-off unit "n"	41.45	dBmV	
Loss to tap-off unit "o" (50' of RG-59/U)	—3.0	dB	
Level at tap-off unit "o"	38.45	dBmV	
(Use white tap, 30 dB isolation)			8.45 dBmV
Feed-thru loss of white tap	—0.25	dB	
Level out of tap-off unit "o"	38.20	dBmV	
Loss to tap-off unit "p" (50' of RG-59/U)	—3.0	dB	
Level at tap-off unit "p"	35.2	dBmV	
(Use white tap, 30 dB isolation)			5.2 dBmV
Feed-thru loss of white tap	—0.25	dB	
Level out of tap-off unit "p"	34.95	dBmV	
Loss to tap-off unit "q" (50' of RG-59/U)	3.0	dB	
Level at tap-off unit "q"	31.95	dBmV	
(Use white tap, 30 dB isolation)			1.95 dBmV
Feed-thru loss of white tap	—0.25	dB	
Level out of tap-off unit "q"	31.7	dBmV	
Loss to tap-off unit "r" (50' of RG-59/U)	—3.0	dB	
Level at tap-off unit "r"	28.7	dBmV	
(Use red tap, isolation loss 20 dB)			8.7 dBmV
Feed-thru loss of red tap	0.5	dB	
	28.3	dBmV	
Et cetera to line end.			

Losses, Leg s		Losses, Leg m	
Isolation	11.5 dB	Isolation	11.5 dB
Feed-through	4.0 dB	Feed-through	6.4 dB
Cable	18.6 dB	Cable	13.2 dB
Total loss	34.1 dB	Total loss	31.1 dB

Since the greatest loss occurs in leg s, adequate amplification must be provided to compensate for this loss. Next, tracing back from the beginning of leg s in Figure 18-1, a splitter and additional cable are encountered. A typical splitter has a loss of 3.5 dB, and each output will have a loss of 3.5 dB with respect to the input signal level. Since 80 feet of cable is also employed, the listing becomes as follows:

Overall System Loss, Leg s	
Isolation loss	11.5 dB
Feed-through loss	4.0 dB
Cable loss, 390 feet	23.4 dB
Splitter loss	3.5 dB
Total loss	42.4 dB

This total loss must be compensated by means of an amplifier with a gain of approximately 43 dB at the required output level. To determine what output level is required, we start with the TV receiver. Approximately 1,000 microvolts in a 75-ohm cable is adequate, and this may be called the 0 dB level. In this case, we write 0 dBmV to denote our reference level, which is 1,000 microvolts across 75 ohms. If we assign this value as the required level at the last receiver on leg s in Figure 18-1, we can calculate back to the amplifier output. The following listing shows how this is done, adding the splitter loss and the extra 20 feet of cable between splitter and amplifier (c and d):

Isolation	11.5 dB
Feed-through	4.0 dB
Cable loss, 330 feet	19.8 dB
Splitter	3.5 dB
	38.8 dB

That is, the loss between the amplifier and the last receiver is 38.8 dB. The level at the receiver must be at least 0 dBmV, and the amplifier output must be 38.8 dBmV. Specifications for three typical MATV amplifiers are shown in Table 18-5. Each of them meets the output requirement in this example. However, the gain requirement of 43 dB is met only by the Model 2880. However, in practice, the Model 3660 with 40 dB gain would be acceptable. To compensate for the lower gain of the Model 3660, the installer could choose an antenna with about 4 dB more gain. System loss can also be reduced to some extent by choosing a low-loss type of lead-in to connect the antenna to the amplifier.

The next step in planning is to determine the color codes of the tap-off units. The calculations are made as before, working backwards step by step. With reference to Figure 18-1, if the amplifier output is 50 dBmV, the listing can be compiled as shown in Table 18-6, calculating the levels remaining at each point in the system. Thus, the level of the signal at n is 41.7 dBmV. This permits the use of a "white" tap (Table 18-2), with an isolation of 30 dB, leaving 11.7 dB or about 4,000 microvolts at the receiver. The "white" tap-off unit has a 0.25 dB feed-through loss which subtracts from the line level, leaving the line at 41.45 dBmV. We continue with this procedure until we come to a level where the use of a "white" tap would put the receiver below 0 dBmV, at which point a "red" tap will be used, and so on.

18–3 INSTALLATION REQUIREMENTS

Where coaxial cable is exposed to lightning or to accidental contact with lightning-arrestor conductors or power conductors operating at a potential greater than 300 volts, the outer conductive shield of the coaxial cable must be grounded at the building premises as close to the point of cable entry as practicable. When the outer conductive shield of a coaxial cable is grounded, a lightning arrestor is not required. A lead-in cable must have a clearance of at least 12 inches from light and power service drops. A cable passing over a building must be installed at least 8 feet above any roof that is accessible for pedestrian traffic. Concealed wiring is generally employed when installing coaxial cable, and the cable must be separated at least 4 inches from any light or power conductors that are not in conduit. However, a coaxial cable may be installed less than 4 inches from such conductors, provided an insulating separator is used between the two systems.

Coaxial cable should not be installed nearer than 6 feet to any lightning conductor. Drip loops must be provided where coaxial cable enters a building. Coaxial cable must not be installed in any raceway, compartment, outlet box, junction box, or other enclosures with conductors for light and power circuits unless the conductors of the different systems are separated by a permanent partition. Coaxial cable may be installed in the same shaft with conductors for light and power, provided the conductors of the two systems are separated at least 2 inches, or where the conductors of either system are encased in noncombustible tubing. Where the lighting or power conductors are installed in a raceway or in metal-sheathed or metal-clad or nonmetallic-sheathed or Type UF cables, neither the 2-inch separation nor the noncombustible tubing is required.

18–4 CATV SYSTEMS

A CATV system is more elaborate than an MATV system, in that a CATV system serves an entire community in a fringe or far-fringe area. Thus, it may serve an entire town in an isolated area. The antenna site is carefully chosen, and is typically on top of a hill or mountain near the town. In most installations, separate high-gain and properly oriented antennas for each active channel are mounted on the antenna tower. Figure 18-5 shows the plan of a typical CATV system.

The head end in a CATV system (Figure 18-2) includes booster amplifiers for the various antennas and a combiner or mixer for the output signals from the amplifiers. In case UHF reception is provided in addition to VHF reception, a translator is utilized for each UHF channel. A translator is a frequency converter that changes a UHF signal frequency into a VHF signal frequency. This process is required because a CATV system employs long runs of coxial cable, and cable losses increase rapidly with an increase in frequency, as noted previously. Therefore, if an 800-MHz signal is "beat down" to 100 MHz, for example, it can be routed through coaxial cable with comparatively little attenuation.

As shown in Table 18-1, the FM band of frequencies is also commonly processed by a CATV system. From the combiner network, the various signal voltages are fed into a trunk cable or sometimes several trunk cables. A trunk cable serves to conduct the signal voltages from their point of origin, such as atop a mountain, to the site of utilization, such as a residential area several miles away. Because the signals are progressively attenuated through a trunk line, amplifiers must be inserted at intervals. An amplification of 50 dB per mile is typically provided. In most installations trunk lines are routed from the head end to the beginning of the distribution system on poles. However, there is a trend toward running the cables underground in ducts or simply in trenches.

When a line amplifier is to be inserted, a buried cable is brought above ground and is routed through a pedestal that houses the amplifier. It is standard practice to utilize 75-ohm coaxial cables and to maintain 75-ohm input and output impedances throughout a CATV system. The distribution system starts with a bridger or bridging amplifier. A simple bridger consists of a passive network comprising a VHF transformer, capacitor, and resistors. It divides the input signal into several portions for supplying feeder lines. A 75-ohm impedance is maintained at all terminals of a bridger. Because a passive device introduces substantial signal attenuation, a bridging amplifier is often employed instead, so that no insertion loss occurs.

When the distribution system is extensive, feeder amplifiers (also called line extenders) may be installed at intervals to maintain an adequate signal level. These are wide-band VHF amplifiers, similar to trunk amplifiers,

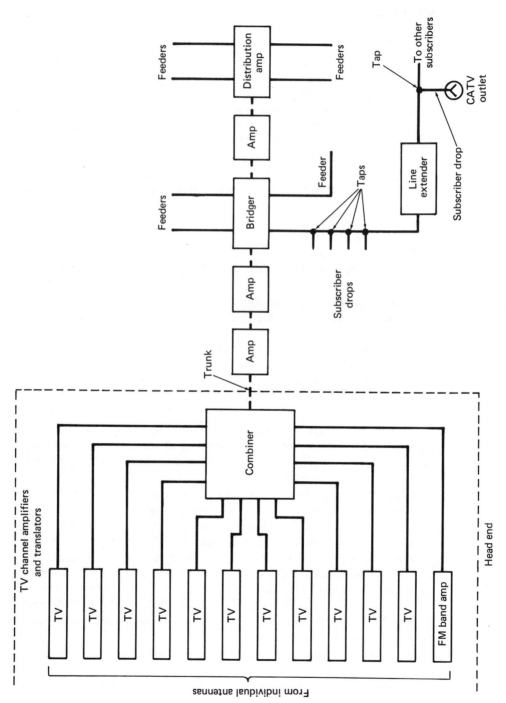

Figure 18-2 Plan of a typical CATV system

that provide up to 25 dB gain over the frequency range from 50 to 220 MHz. All amplifiers or signal splitters in a CATV system (Figure 18-2) are usually provided with a tilt control or equalizer. This is a bandpass type of filter device that can be adjusted to attenuate the low end or the high end of the VHF band as required. In most situations the high end of the band will have become attenuated, so that the tilt control is adjusted to introduce a compensating attenuation at the low end of the band.

Feeder cables and cable installations inside a building are also called distribution cables. In the strict sense of the term, distribution cables are tapped off from a feeder cable. A tap point is called a subscriber tap. A bridging amplifier always has at least one trunk output in addition to the feeder outputs. Thus, a trunk line continues through a bridger. The last amplifier at the end of a trunk line is called a distribution amplifier. It is basically the same as a bridging amplifier, except that the trunk line terminates at the distribution amplifier. A feeder line can be run at distances up to 1,000 feet from a bridger. Beyond 1,000 feet, a line extender is required to maintain an adequate signal level.

A subscriber drop is a cable that is tapped at some point along a feeder line. Coupling is provided by the tap in a manner that does not introduce objectionable disturbance into the feeder cable. The simplest form of tap coupler is an isolating resistor connected between the tap point and the cable to the TV receiver. Another form of tap coupler is a small capacitor. Capacitive coupling is sometimes preferred because it introduces a tilt that tends to compensate for the frequency slope of the subscriber drop cable. The signal level applied to a TV receiver from a CATV system is typically 1,500 microvolts. Receivers are generally connected into CATV and MATV circuits via wall outlets.

18-5 LIGHTNING ARRESTORS

Any television antenna must be protected from lightning to pass inspection. The NEC requires that the metal sheath of antenna cables that enter buildings which are likely to contact electric light or power conductors shall be grounded or shall be interrupted close to the entrance of the building by an insulating joint or equivalent device. A grounding conductor must not be smaller than No. 18 gage. Masts and metal structures supporting antennas must be grounded. Each conductor of a lead-in from an antenna must be provided with a lightning arrestor, unless coaxial cable is utilized. Lightning arrestors must be located outside the building, or inside the building between the point of entrance of the lead-in and the receiver, and as near as practicable to the entrance of the conductors to the building.

Any interference eliminators connected to the power-supply leads must be of a type approved for the purpose, and they must not be exposed to physical damage. If the power line is utilized as an antenna, the device for coupling the receiver to the supply circuit must be specially approved for the purpose. Indoor antennas do not require lightning protection; however, the lead-in from either an indoor antenna or an outdoor antenna must not be run nearer than two inches to conductors of other wiring systems in the premises.

QUESTIONS

1. Where is an MATV system generally used?

2. What signals are usually transmitted by an MATV and a CATV system?

3. What is the simplest method of isolation between each receiver and the line?

4. What is the impedance of most cables in an MATV system?

5. Where are isolation losses developed?

6. Where are higher isolation losses developed?

7. What is the minimum distance that a lead-in cable may be installed near a power service drop?

8. What is the name of the loop that is placed in a coaxial cable as it enters a building?

9. What is a translator?

10. What is a trunk cable?

11. What type of standard cable is used in a CATV system?

12. What is another name for feeder cables?

13. What is the name of the output from a bridging amplifier?

14. What is the simplest form of tap coupler?

15. What is the typical signal level applied to a receiver from a CATV system?

19 TELEPHONE SYSTEM WIRING

19–1 GENERAL CONSIDERATIONS

Telephone wiring systems utilize insulated conductors such as shown in Figure 19-1. Rubber insulation is widely used with an outer covering of fibrous or thermoplastic material. Since telephone currents are small, compared with lighting or power currents, conductor sizes are correspondingly small. In many cases, the conductor size is sufficiently small that it will melt in case of an accidental surge of substantial current. For example, No. 24 wire may be employed. In turn, the NEC does not require fuses to be installed in the telephone circuit. On the other hand, lightning arrestors are always required. In rural areas, bare wire may be installed for telephone lines on pole-mounted insulators. Most installations, however, employ lead-sheathed multiconductor cables suspended from poles or buried underground.

When telephone cables or lines are installed on poles, they are run below any light or power conductors that may be on the same pole, as depicted in Figure 19-2. The NEC does not permit telephone conductors to be installed on the same crossarm as light or power conductors. Note that the insulators on the telephone crossarm must provide a climbing space of at least 24 inches, if the power conductors operate under 300 volts. If the power conductors operate at more than 300 volts, at least 30 inches of climbing space must be provided. In case only a telephone crossarm is installed, with no light or power crossarm above, no climbing space need be provided on the telephone crossarm.

The entrance cable (which may be termed a telephone service drop) must have at least 8 feet clearance over any roofs below the cable, in case the roof might be walked upon. On the other hand, if there is no normal access to a roof, such as a residential garage, there is no required clearance.

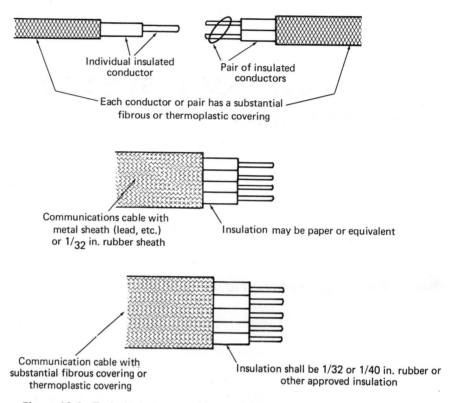

Individual insulated conductor

Pair of insulated conductors

Each conductor or pair has a substantial fibrous or thermoplastic covering

Communications cable with metal sheath (lead, etc.) or 1/$_{32}$ in. rubber sheath

Insulation may be paper or equivalent

Communication cable with substantial fibrous covering or thermoplastic covering

Insulation shall be 1/32 or 1/40 in. rubber or other approved insulation

Figure 19-1 Typical telephone-system conductors

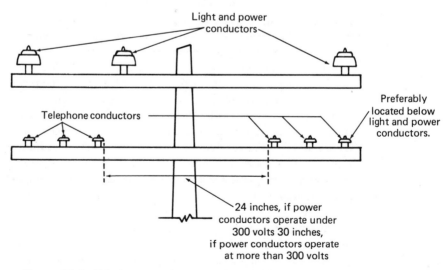

Light and power conductors

Telephone conductors

Preferably located below light and power conductors.

24 inches, if power conductors operate under 300 volts 30 inches, if power conductors operate at more than 300 volts

Figure 19-2 Telephone conductors are installed on a separate crossarm, preferably below light or power conductor crossarms

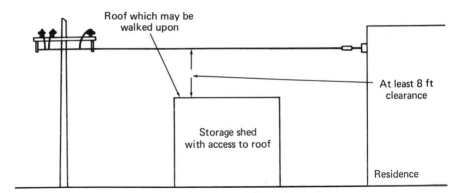

Figure 19-3 An installation that requires an 8-foot clearance

Figure 19-4 Installation of a service-drop grip clamp

Figure 19-3 depicts a situation in which an 8-foot clearance must be maintained. The entrance cable or service drop is commonly installed with a grip clamp, as illustrated in Figure 19-4. The grip clamp has a wedge construction which binds the cable more tightly into the device as the tension increases. This arrangement simplifies the installation, inasmuch as a service rack is dispensed with.

19–2 LIGHTNING AND SURGE PROTECTION

A protector is installed as near as practical to the point where the telephone conductors enter the building, as shown in Figure 19-5. The protector is usually mounted on the outside wall although it may be mounted inside the building, if necessary. A protector is basically a lightning arrestor, consisting of carbon rods spaced 0.005 inch from grounded metal blocks. Any substantial voltage surge will arc across the narrow gap and the surge will be conducted harmlessly to ground. Figure 19-6 illustrates a telephone protector and shows the circuit that is employed. Note that the fuses are often omitted when small-gage conductors are utilized.

The grounding conductor for a telephone protector is No. 18 copper wire or larger, run as directly as possible to a water pipe or made ground. It must have rubber insulation or approved thermoplastic insulation at least 1/32 inch in thickness. If the ground wire is subject to possible mechanical dam-

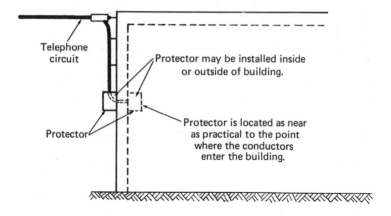

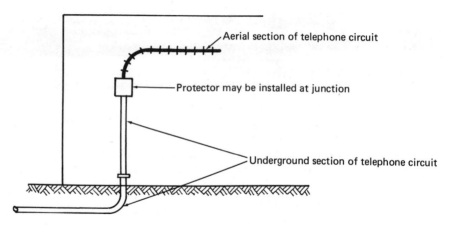

Figure 19-5 Installation of telephone circuit protectors

age, it must be enclosed in conduit, conventional pipe, EMT, or equivalent. A ground connection is preferably made to a cold-water pipe. However, a continuous and extensive underground gas-piping system may be used, or some effectively grounded metallic structure, if necessary. A made ground, consisting of a ground rod or pipe driven into permanently damp soil, is also permissible. A protector must not be grounded to hot-water pipes, steam pipes, or lightning-rod conductors.

However, if there are several made grounds in an electrical system, the telephone-protector ground rod may be bonded together with other ground rods, as shown in ,Figure 19-7. Also, if a power system has several grounds, the telephone-protector ground wire may be connected to one of the power-system grounds, as shown in Figure 19-8. Note that when a gas-pipe system

(a)

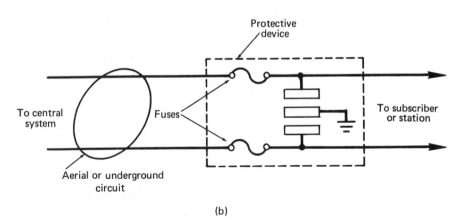

(b)

Figure 19-6 Telephone protective arrangement (a) appearance of protector (Courtesy Winegard Antenna Systems) (b) circuit configuration

is used for grounding a protector, the ground connection must be made between the gas meter and the street mains. This precaution avoids the possibility of a poor ground circuit through the meter housing. When the protector is installed inside a building and conduit is used for the entrance conductors, the conduit must be grounded as shown in Figure 19-9.

The NEC requires an approved protector on each circuit that is run partly or entirely in aerial wire or aerial cable that is not limited to one block. A protector must also be provided on each circuit (aerial or underground),

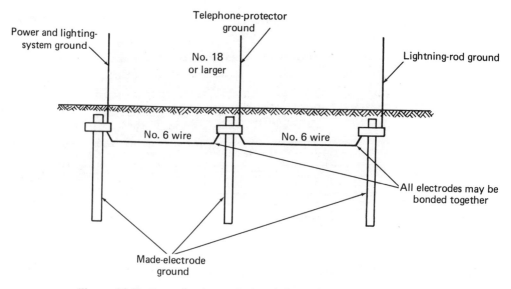

Figure 19-7   Ground rods may be bonded together

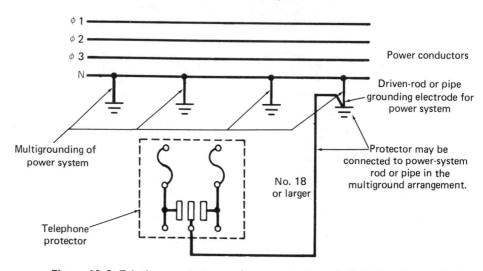

Figure 19-8   Telephone protector may be connected to an individual multiground point

located within the block containing the building served which could be exposed to accidental contact with light and power conductors that operate at a potential greater than 300 volts. The word "block" as used by the NEC means a square or portion of a city, town, or village that is enclosed by streets. It includes the alleys so enclosed, but not any street. Figure 19-10 clarifies this definition of "block." Here the word "exposed" means that the circuit is so arranged that contact with another circuit may result if supports or insulation ever fail.

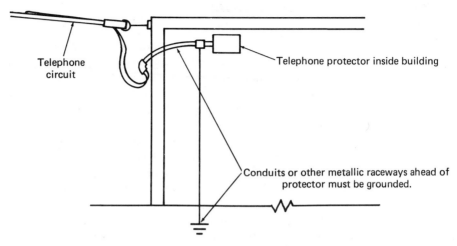

Telephone circuit

Telephone protector inside building

Conduits or other metallic raceways ahead of protector must be grounded.

Figure 19-9 Conduit for entrance conductors must be grounded

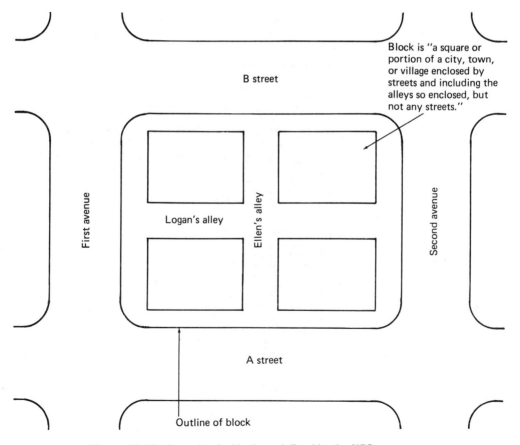

B street

Block is "a square or portion of a city, town, or village enclosed by streets and including the alleys so enclosed, but not any streets."

First avenue

Second avenue

Logan's alley

Ellen's alley

A street

Outline of block

Figure 19-10 Example of a block, as defined by the NEC

19–3 GROUPING OF CONDUCTORS

More than one conductor of a communication system may be installed in a raceway, but all communication conductors must be partitioned from light and power conductors, as shown in Figure 19-11. A telephone cable or wires must be installed at least 4 inches away from open conductors used for light or power, as illustrated in Figure 19-12(a). On the other hand, when light or power conductors are enclosed by conduit, no minimum spacing is required for telephone conductors, as indicated in Figure 19-12(b). A special case is encountered in the installation of elevator-car telephones, as depicted in Figure 19-13. The telephone conductors may be run in the traveling cable with the power conductors, provided all of the conductors are insulated for the maximum voltage existing in the cable.

When telephone cables are installed underground with light or power cables, they may occupy the same manhole or handhole, provided the telephone conductors are in a section separated from the light or power conduc-

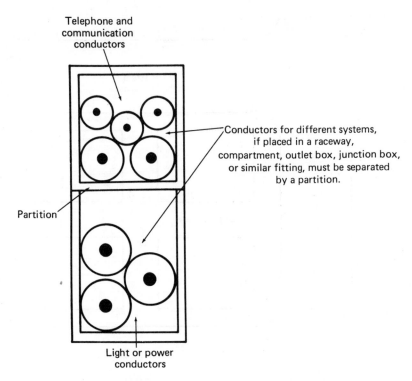

Figure 19-11   Installation of communication and light or power conductors in a raceway

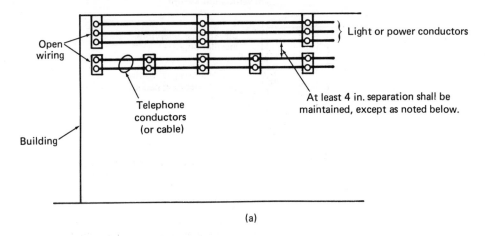

(a)

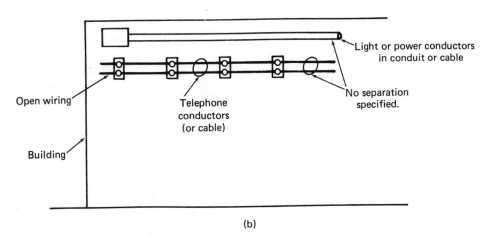

(b)

Figure 19-12 Spacing required between telephone and light-power conductors (a) four-inch separation required from open conductors (b) no minimum separation specified from conduit

tors by means of brick, concrete, or tile partitions. Figure 19-14 depicts a typical installation with a telephone duct and a light/power duct entering the same handhole that is constructed of concrete with a concrete partition between the two systems. This is essentially a combination junction point where ducts change direction. Conductors for both systems are pulled through their associated ducts at the time of installation, and connected together in the handhole.

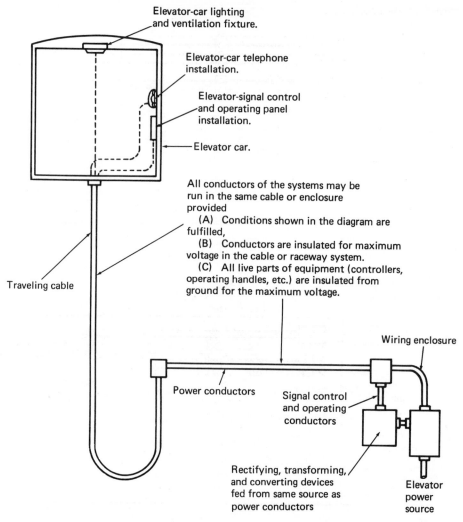

Elevator-car lighting and ventilation fixture.

Elevator-car telephone installation.

Elevator-signal control and operating panel installation.

Elevator car.

All conductors of the systems may be run in the same cable or enclosure provided

(A) Conditions shown in the diagram are fulfilled,

(B) Conductors are insulated for maximum voltage in the cable or raceway system.

(C) All live parts of equipment (controllers, operating handles, etc.) are insulated from ground for the maximum voltage.

Traveling cable

Wiring enclosure

Power conductors

Signal control and operating conductors

Rectifying, transforming, and converting devices fed from same source as power conductors

Elevator power source

Figure 19-13 Installation of telephone conductors in the traveling cable of an elevator

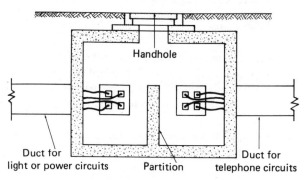

Handhole

Duct for light or power circuits

Partition

Duct for telephone circuits

Figure 19-14 Installation of telephone cables with light/power cables in a duct-and-handhole arrangement

19–4 CONNECTIONS TO HANDSET HOLDERS

Figure 19-15 shows how a telephone cable is connected to the terminal block in a handset holder. Three conductors are usually employed, called Line 1 (L1), Line 2 (L2), and common ground (G). However, the cable may have four conductors, one of which is left unconnected in this situation. Conductors are color-coded so that the electrician can identify each conductor at the start and at the end of a run. Exposed wiring is in very wide use because residences are often built with no thought of telephone wiring installation. An exposed run of cable is secured by means of staples. However, when a residence or other building is planned by an architect, the construction usually provides for concealed telephone wiring which is installed in the same basic manner as concealed light and power wiring. Baseboard outlets may be provided so that a handset can be moved conveniently from one room to another.

Figure 19-15 Connection of a telephone cable to the terminal block in a handset holder

QUESTIONS

1. How do telephone currents compare to lighting circuit currents?
2. What safety requirement is necessary on telephone circuits?
3. What conductors may not be installed on the same crossarm as telephone conductors?

4.   What is the voltage level at which a 30-inch climbing space must be provided between insulators on a crossarm?

5.   What is the minimum clearance that a telephone service drop must have above a roof?

6.   Where should the grounding conductor of a telephone service be connected?

7.   When a gas line is used for the grounding circuit, where should the cable be connected to the pipe?

REVIEW QUESTIONS FOR PART III

1.   Is galvanized-iron conduit used in underground service entrances for large multiple-occupancy buildings?

2.   Why may a multiple-occupancy building require more than one service?

3.   Must each occupant in a multiple-occupancy building have access to his disconnect switch?

4.   In a three-phase service consisting of a four-wire installation with a neutral wire, what is the voltage between any two of the phase wires, and what is the voltage between any phase wire and the neutral wire?

5.   What is a demand factor?

6.   How many groups of hazardous locations does the NEC define?

7.   Does a three-phase circuit require larger gage wires than an equivalent single-phase circuit?

8.   How many wires are used in a 120/208-volt three-phase system?

9.   Explain what is meant by phase rotation, or phase sequence.

10.   When must an explosive-proof panelboard be installed?

11.   How many watts are equal to 1 horsepower?

12.   What is the typical efficiency of a 1-horsepower motor at full load?

13.   How is the housing of a motor grounded?

14.   Explain the difference between an induction motor and a synchronous motor.

15.   What is a typical ratio of starting current to full-load current for a motor?

16.   When does the greatest current demand occur in an automotive wiring system?

17.   How is a circuit traced through an automotive wiring diagram?

18.   What is meant by a negative-ground installation?

19.   Explain what is meant by a blind circuit.

20.   When is a marine electrical wiring installation started?

21.   How are electric welding arcs powered in a marine wiring project?

22.   Explain the danger of a "flash" from an arc-welding unit.

23.   What is meant by a high-cycle system?

24.   Is high-cycle current less dangerous than low-cycle current?

25.   What is a kick pipe?

26.   Does an aircraft electrical system use AC power or DC power?

27. Explain what is meant by a bond in an aircraft electrical system.
28. What is the frequency range of aircraft inverters?
29. Where are recognition lights usually located?
30. What is the basic function of a CATV or MATV system?
31. How are UHF signals processed in a CATV system?
32. Explain the main sections of a distribution system.
33. What is the standard value for cable impedance in CATV systems?
34. How many dB amplification per mile is generally provided in a CATV system?
35. What is the typical signal level applied to a receiver from a CATV system?
36. How do current levels in telephone systems compare with those in lighting systems?
37. What is the minimum clearance that a telephone service drop must have above a roof?
38. Is telephone wiring always concealed inside of a building?
39. May wires of a communication system be installed with light and power wires in a raceway?
40. How is a telephone wiring system protected against lightning?

GLOSSARY

AC abbreviation for alternating current

Access Fitting a fitting that permits access to conductors in concealed or enclosed wiring, elsewhere than at an outlet

Air-Blast Transformer a transformer cooled by forced circulation of air through its core and coils

Air Circuit Breaker a circuit breaker in which the interruption occurs in air

Air Switch a switch in which the interruption of the circuit occurs in air

Alive electrically connected to a source of emf, or electrically charged with a potential different from that of the earth. Also: Practical synonym for "current-carrying"

Alternating Current a periodic current, the average value of which over a period is zero

Alternator (Synchronous Generator) a synchronous alternating-current machine that changes mechanical power into electrical power

Aluminum Conductor a conductor made wholly of aluminum

Ammeter an instrument for measuring electric current

Ampacity current rating in amperes, as of a conductor

Ampere a charge flow of one coulomb per second

Annunciator an electromagnetically operated signaling apparatus that indicates whether a current is flowing or has flowed in one or more circuits

Apparent Power in a single-phase, two-wire circuit, the product of the effective current in one conductor multiplied by the effective voltage between the two points of entry

Appliance current-consuming equipment, fixed or portable, such as heating or motor-operated equipment

Appliance Branch Circuits circuits supplying energy either to permanently wired appliances or to attachment-plug receptacles—that is, appliance or convenience outlets—or to a combination of permanently wired appliances and additional attachment-plug outlets on the same circuit, such circuits to have no permanently connected lighting fixtures

Appliance, Portable an appliance that can be carried readily from place to place and that can be served by means of a flexible extension cord and attachment plug

Arcing Contacts contacts on which an arc is drawn after the main contacts of a switch or circuit-breaker have parted

Arcing Time of Fuse the time elapsing from the severance of the fuse link to the final interruption of the circuit under specified conditions

Arc-over of Insulator a discharge of power current in the form of an arc, following a surface discharge over an insulator

Armor Clamp a fitting for gripping the armor of a cable at the point where the armor terminates, or where the cable enters a junction box or other apparatus

Armored Cable a cable provided with a wrapping of metal, usually steel wires, primarily for the purpose of mechanical protection

Arrestor, Lightning a device that reduces the voltage of a surge applied to its terminals, and restores itself to its original operating condition

Autotransformer a transformer in which part of the winding is common to both the primary and secondary circuits

B&S Wire Gage Brown and Sharpe wire gage; same as AWG

Back-Connected Switch a switch in which the current-carrying conductors are connected to studs in back of the mounting base

Bank an assemblage of fixed contacts in a rigid unit over which wipers or brushes may move and make connection with the contacts

Bank, Duct an arrangement of conduit which provides one or more continuous ducts between two points

Benchboard a switchboard with a horizontal section for control switches, indicating lamps, and instrument switches; may also have a vertical instrument section

Bidirectional Current a current that has both positive and negative values

Bond, Cable an electric connection across a joint in the armor or lead sheath of a cable, or between the armor or

sheath to ground, or between the armor or sheath of adjacent cable

Bond Wire a bare grounding wire that runs inside of an armored cable

Box, Conduit a metal box adapted for connection to conduit for installation of wiring, making connections, or mounting devices

Box, Cutout an enclosure designed for surface mounting and having swinging doors or covers secured directly to and telescoping with the walls of the box proper

Box, Equipment a metal box designed for housing electrical equipment and provided with one or more doors attached directly to the box, or attached to a trim or cover which does not surround the door or doors

Box, Junction an enclosed distribution panel for connection or branching of one or more electric circuits without making permanent splices

Box, Junction (Interior Wiring) a metal box with blank cover for joining runs of conduit, electrical metallic tubing, wireway, or raceway, and providing space for connection and branching of enclosed conductors

Box, Pull a metal box with a blank cover that is used in a run of conduit, etc. to facilitate pulling in the conductors; it may also be installed at the end of one or more conduit runs for distribution of the conductors

Branch Circuit that portion of a wiring system extending beyond the final automatic overload protective device

Branch Circuit, Appliance a circuit supplying energy either to permanently wired appliances or to attachment-plug receptacles such as appliance or convenience outlets, and having no permanently connected lighting fixtures

Branch Circuit Distribution Center a distribution circuit at which branch circuits are supplied

Branch Circuit, Lighting a circuit that supplies energy to lighting outlets only

Branch Conductor a conductor that branches off at an angle from a continuous run of conductor

Branch Joint a multiple joint for connection of a branch conductor or cable to a main conductor or cable, wherein the latter continues beyond the branch

Break the break of a circuit-opening device is the minimum distance between the stationary and movable contacts when the device is in its open position

Breakdown also termed "puncture," denoting a disruptive discharge through insulation

Breaker, Line a device that combines the functions of a contactor and a circuit-breaker

Buried Cable a cable installed under the surface of the soil in such manner that it cannot be removed without digging up the soil

Bus a conductor or group of conductors that serve as a common connection for three or more circuits in a switchgear assembly

Bushing also termed "insulating bushing"; a lining for a hole for insulation and/or protection from abrasion of one or more conductors passing through it

BX Cable cable with galvanized steel spiral armor

Cabinet an enclosure for either surface or flush mounting, provided with a frame, mat, or trim

Cable Fault a partial or total local failure in the insulation or continuity of the conductor

Cable Joint also termed a "splice"; a connection between two or more individual lengths of cables, with their conductors individually connected, and with protecting sheaths over the joint

Cable, Service service conductors arranged in the form of a cable

Cable Sheath the protective covering, such as lead, applied over a cable

Charge, Electric an inequality of positive and negative electricity in or on a body. The charge stored in a capacitor (condenser) corresponds to a deficiency of free electrons on the positive plate, and to an excess of free electrons on the negative plate

Choke Coil a low-resistance coil with sufficient inductance to substantially impede AC or transient currents

Circuit, Electric a conducting path through which electric charges may flow. A DC circuit is a closed path for charge flow; an AC circuit is not necessarily closed and may conduct in part by means of an electric field (displacement current)

Circuit, Earth (Ground) Return an electric circuit in which the ground serves to complete a path for charge flow

Circuit, Magnetic a closed path for establishment of magnetic flux (magnetic field) that has the direction of the magnetic induction at every point

Cleat an assembly of a pair of insulating material members with grooves for holding one or more conductors at a definite distance from the mounting surface

Clip, Fuse contacts on a fuse support for connecting a fuseholder into a circuit

Closed-Circuit Voltage the terminal voltage of a source of electricity under a specified current demand

Closed Electric Circuit a continuous path or paths providing for charge flow. In an AC closed circuit, charge flow may be changed into displacement current "through" a capacitor (condenser)

Coil a conductor arrangement (basically a helix or spiral) that concentrates the magnetic field produced by electric charge flow

Coil, Inductance (Inductor) a device, the primary purpose of which is to introduce inductance into an electric circuit

Composite Conductor a conductor consisting of two or more strands of different metals, operated in parallel

Concealed to be made inaccessible by the structure or finish of a building; also, wires run in a concealed raceway

Condenser also termed "capacitor"; a device that stores electric charge by means of an electric field

Conductor a substance that has free electrons or other charge carriers that permit charge flow when an emf is applied across the substance

Conduit a structure containing one or more ducts; commonly formed from iron pipe or electrical metallic tubing

Conduit Fittings accessories used to complete a conduit system, such as boxes, bushings, and access fittings

Conduit, Flexible Metal a flexible raceway of circular form for enclosing wires or cables; usually made of steel wound helically and with interlocking edges, and with a weather-resistant coating

Conduit, Rigid Steel a raceway made of mild steel pipe with a weather-resistant coating

Conduit Run a duct bank; an arrangement of conduit with a continuous duct between two points in an electrical installation

Contactor an electric power switch, not operated manually, and designed for frequent operation

Contacts conducting parts that employ a junction that is opened or closed to interrupt or complete a circuit

Control Relay a relay used to initiate or permit a predetermined operation in a control circuit

Control Switch a switch for controlling electrically operated devices

Coulomb an electric charge of 6.28 $\times 10^{18}$ electrons. One coulomb is transferred when a current of 1 ampere continues past a point for 1 second

Current the rate of charge flow. A current of 1 ampere is equal to a flow rate of 1 coulomb per second

Cycle the complete series of values that occur during one period of a periodic quantity. The unit of frequency, the hertz, is equal to one cycle per second

DC abbreviation for direct current

Dead functionally conducting parts of an electrical system that have no potential difference or charge (voltage of zero with respect to ground)

Demand Factor the ratio of the maximum demand of a system, or part of a system, to the total connected load of the system

Diagram, Connection a drawing showing the connections and interrelations of devices employed in an electrical circuit

Dielectric a medium or substance in which a potential difference establishes an electric field that is subsequently recoverable as electric energy

Direct Current a unidirectional current with a constant value. "Constant value" is defined in practice as a value that has negligible variaton

Direct EMF also termed "direct voltage"; an emf that does not change in polarity and has a constant value (one of negligible variation)

Discharge an energy conversion involving electrical energy. Examples: discharge of a storage battery; discharge of a capacitor; lightning discharge of a thundercloud

Disruptive Discharge a rapid and large current increase through an insulator due to insulation failure

Distribution Center a point of installation for automatic overload protective devices connected to buses where an electrical supply is subdivided into feeders and/or branch circuits

Drop, Voltage an IR voltage between two specified points in an electric circuit

Duct a single enclosed runway for conductors or cables

E symbol for voltage

Edison Base the screw-type base used on ordinary incandescent lamps

Effective Value the effective value of a sine-wave AC current or voltage is equal to 0.707 of peak. Also called the root-mean-square (rms) value, it produces the same I^2R power as an equal DC value

Effectively   Grounded   this means grounded by means of a ground connection of sufficiently low impedance that fault grounds which may occur cannot build up voltages dangerous to connected equipment

Efficiency   the ratio of output power to input power, usually expressed as a percentage

Electrical Units   in the practical system, electrical units comprise the volt, the ampere, the ohm, the watt, the watt-hour, the coulomb, the mho, the henry, the farad, the joule, and the hertz

Electricity   a physical entity associated with the atomic structure of matter that occurs in polar forms (positive and negative) which are separable by expenditure of energy

Electrode   a conducting substance through which electric current enters or leaves in devices that provide electrical control or energy conversion

Electromagnetic Induction   a process of generation of emf by movement of magnetic flux which cuts an electrical conductor

Electromotive   Force   (EMF)   an energy-charge relation that results in electric pressure which produces or tends to produce charge flow

Electron   the subatomic unit of negative electricity. It is a charge of 1.6 $\times$ 10^{-19} coulomb

Electronics   the science treating of charge flow in vacuum, gases, and crystal lattices

Electroplating   the electrical deposition of metallic ions as neutral atoms on an electrode immersed in an electrolyte

Electrostatics   a branch of electrical science dealing with the laws of electricity at rest

EMT   abbreviation for electrical metallic tubing (thin-wall conduit)

Energy   the amount of physical work that a system is capable of doing. Electrical energy is measured in watt-seconds, or the product of power and time

Entrance Cap   a service head, or weather head

Entrance, Duct   an opening of a duct at a distribution box or other accessible location

Equipment, Service   a circuit-breaker or switches and fuses with their accessories, installed near the point of entry of service conductors to a building

Exciter   an auxiliary generator for supplying electrical energy to the field of another electrical machine

Face Plate   a cover for a switch or receptacle box; also called a wall plate

Farad   a unit of capacitance that is defined by the production of 1 volt across the capacitor terminals when a charge of 1 coulomb is stored

Fault Current   an abnormal current flowing between conductors or from a conductor to ground due to an insulation defect, arc-over, or incorrect connection

Feeder   a conductor or a group of conductors for connection of generating stations, substations, generating and substations, or a substation and a feeding point

Filament   a wire or ribbon of conducting (resistive) material that develops light and heat energy due to electric charge flow; light radiation is also accompanied by electron emission

Fixture Stud   a fitting for mounting a lighting fixture in an outlet box

Flashover   a disruptive electrical discharge around or over (but not through) an insulator

Fluorescence   an electrical discharge process involving radiant energy transferred by phosphors into radiant energy that provides increased luminosity

Flux   electrical field energy distributed in space, in a magnetic substance, or in a dielectric. Flux is commonly represented diagramatically by means of flux lines denoting magnetic or electric forces

Force an elementary physical cause capable of modifying the motion of a mass

Frequency the number of periods occurring in unit time of a periodic process, as in the flow of electric charge

Frequency Meter an instrument that measures the frequency of an alternating current

Fuse a protective device with a fusible element that opens the circuit by melting when subjected to excessive current

Fuse Cutout an assembly consisting of a fuse support and holder; it may also include a fuse link

Fuse Element also termed "fuse link"; the current-carrying part of a fuse which opens the circuit when subjected to excessive current

Fuseholder a supporting device for a fuse that provides terminal connections

Fustat a nontamperable type of fuse

Gap (*Spark Gap*) a high-voltage device with electrodes between which a disruptive discharge of electricity may pass, usually through air. A sphere gap has spherical electrodes; a needle gap has sharply pointed electrodes; a rod gap has rods with flat ends

Ground also termed "earth"; a conductor connected between a circuit and the soil; a chassis-ground is not necessarily at ground potential, but is taken as a zero-volt reference point. An accidental ground occurs due to cable insulation faults, an insulator defect, etc.

Grounding Electrode a conductor buried in the earth, for connection to a circuit. The buried conductor is usually a cold-water pipe, to which connection is made with a ground clamp

Ground Lug a lug for convenient connection of a grounding conductor to a grounding electrode or device to be grounded

Ground Outlet an outlet provided with a polarized receptacle with a grounded contact for connection of a grounding conductor

Ground Switch a switch for connection or disconnection of a grounding conductor

Ground Wire a wire that runs from the service switch to an earth ground

Guy a wire or other mechanical member having one end secured and the other end fastened to a pole or structural part maintained under tension

Hanger also termed "cable rack"; a device usually secured to a wall to provide support for cables

Heat Coil a protective device for opening and/or grounding a circuit by switching action when a fusible element melts due to excessive current

Heater in the strict sense, a heating element for raising the temperature of an indirectly heated cathode in a vacuum or gas tube. Also applied to appliances such as space heaters and radiant heaters

Henry the unit of inductance; it permits current increase at the rate of 1 ampere per second when 1 volt is applied across the inductor terminals

Hickey a fitting for mounting a lighting fixture in an outlet box. Also, a device used with a pipe handle for bending conduit

Horn Gap a form of switch provided with arcing horns for automatically increasing the length of the arc and thereby extinguishing the arc

I symbol for electric current

Identified Terminal in an electrical device, a light-colored or marked terminal, to which the neutral wire is connected

Impedance opposition to AC current by a combination of resistance and reactance; impedance is measured in ohms

Impulse an electric surge of unidirectional polarity

Indoor Transformer a transformer that must be protected from the weather

Induced Current a current that results in a closed conductor due to cutting of lines of magnetic force

Inductance an electrical property of a resistanceless conductor that may have a coil form, and which exhibits inductive reactance to an AC current. All practical inductors also have at least a slight amount of resistance

Inductor a device such as a coil with or without a magnetic core that develops inductance, as distinguished from the inductance of a straight wire

Instantaneous Power the product of an instantaneous voltage by the associated instantaneous current

Instrument an electrical device for measurement of a quantity under observation, or for presenting a characteristic of the quantity

Interconnection, System a connection of two or more power systems

Interconnection Tie a feeder that interconnects a pair of electric supply systems

Interlock an electrical device depending on its operation from another device, for controlling subsequent operations

Internal Resistance the effective resistance connected in series with a source of emf due to resistance of the electrolyte, winding resistance, etc.

IR Drop a potential difference produced by charge flow through a resistance

Isolating Switch an auxiliary switch for isolating an electric circuit from its source of power; it is operated only after the circuit has been opened by other means

Joint an electrical connection of two or more wires; also called a splice

Joule a unit of electrical energy; also called a watt-second; the transfer of 1 watt for 1 second

Joule's Law states that the rate at which electrical energy is changed into heat energy is proportional to the square of the current

Jumper a short length of conductor for making a connection between terminals, around a break in a circuit, or around an electrical instrument

Junction a point in a parallel or series-parallel circuit where current branches off into two or more paths

Junction Box an enclosed distribution panel for the connection or branching of one or more electrical circuits, not using permanent splices. In the case of interior wiring, a junction box consists of a metal box with a blank cover; it is inserted in a run of conduit, raceway, or tubing

Kilowatt a unit of power, equal to 1000 watts

Knockout a scored portion in the wall of a box or cabinet that can be removed easily by striking with a hammer; a circular hole is provided thereby for accommodation of conduit or cable

KVA kilovolt-amperes; the product of volts and amperes divided by 1,000

L symbol for inductance

Lag denotes that a given sine wave passes through its peak at a later time than a reference sine wave

Lampholder also termed "socket" or "lamp receptacle"; a device for mechanical support of and electrical connection to a lamp

Lay the lay of a helical element of a cable is equal to the axial length of a turn

Lead denotes that a given sine wave passes through its peak at an earlier time than a reference sine wave

Leakage, Surface passage of current over the boundary surfaces of an insulator as distinguished from passage of current through its bulk

Leg of a Circuit one of the conductors in a supply circuit between which the maximum supply voltage is maintained

Limit Switch a device that automatically cuts the power off at or near the limit of travel of a mechanical member

Load   the load on an AC machine or apparatus is equal to the product of the rms voltage across its terminals and the rms current demand

Locking Relay   a relay that operates to make some other device inoperative under certain conditions

Loom   see *Tubing, Flexible*

Luminosity   relative quantity of light

Magnet   a body that is the source of a magnetic field

Magnetic Field   the space that contains the energy that is distributed in the vicinity of a magnet and in which magnetic forces are apparent

Meter   a unit of length equal to 39.37 inches; an electrical instrument for measurement of voltage, current, power, energy, phase angle, synchronism, resistance, reactance, impedance, inductance, capacitance, etc.

Mounting, Circuit-Breaker   supporting structure for a circuit breaker

Multiple Feeder   two or more feeders connected in parallel

Multiple Joint   a joint for connecting a branch conductor or cable to a main conductor or cable, to provide a branch circuit

Multiplier, Instrument   a series resistor connected to a meter mechanism so as to provide a higher voltage-indicating range

Mutual Inductance   an inductance that is common to the primary and secondary of a transformer, resulting from primary magnetic flux that cuts the secondary winding

NEC   abbreviation for National Electrical Code

Negative   a value less than zero; an electric polarity sign indicating an excess of electrons at one point with respect to another point; a current sign indicating charge flow away from a junction

Network   a system of interconnected paths for charge flow

No-Load Current   the current demand of a transformer primary, when no current demand is made on the secondary

Normally Closed   denotes the automatic closure of contacts in a relay when deenergized (not applicable to a latching relay)

Normally Open   denotes the automatic opening of contacts in a relay when deenergized (not applicable to a latching relay)

Ohm   the unit of resistance; a resistance of 1 ohm sustains a current of 1 ampere when 1 volt is applied across the resistance

Ohmmeter   an instrument for measuring resistance values

Ohm's Law   states that current is directly proportional to applied voltage, and inversely proportional to resistance, reactance, or impedance

Open-Circuit Voltage   the terminal voltage of a source under conditions of no current demand. The open-circuit voltage has a value equal to the emf of the source

Open-Wire Circuit   a circuit constructed from conductors that are separately supported on insulators

Operator   a low-voltage circuit relay

Outdoor Transformer   a transformer with weatherproof construction

Outlet   a point in a wiring system from which current is taken to supply fixtures, lamps, heaters, etc.

Outlet, Lighting   an outlet used for direct connection of a lampholder, lighting fixture, or a cord that supplies a lampholder

Outlet, Receptacle   an outlet used with one or more receptacles that are not of the screw-shell type

Overload Protection   interruption or reduction of current under conditions of excessive demand, provided by a protective device

Ozone   a compound consisting of three atoms of oxygen, produced by the action of electric sparks, or special electrical devices

Panelboard a distribution point where an incoming set of wires branches into various other circuits

Peak Current the maximum value (crest value) of an alternating current

Peak Voltage the maximum value (crest value) of an alternating voltage

Peak-to-Peak Value the value of an AC waveform from its positive peak to its negative peak. In the case of a sine wave, the peak-to-peak value is double the peak value

Pendant a fitting that is suspended from overhead by a flexible cord which may also provide electrical connection to the fitting

Pendant, Rise-and-Fall a pendant that can be adjusted in height by means of a cord adjuster

Period the time required for an AC waveform to complete one cycle

Permanent Magnet a magnetized substance that has substantial retentivity

Phase the time of occurrence of the peak value of an AC waveform with respect to the time of occurrence of the peak value of a reference waveform. Phase is usually stated as the fractional part of a period

Phase Angle an angular expression of phase difference. It is commonly expressed in degrees and is equal to the phase multiplied by 360°

Plug a device that is inserted into a receptacle for connection of a cord to the conductor terminations in the receptacle

Polarity an electrical characteristic of emf that determines the direction in which current tends to flow

Pole the pole of a magnet is an area at which its flux lines tend to converge or diverge

Positive a value greater than zero; an electric polarity sign denoting a deficiency of electrons at one point with respect to another point; a current sign indicating charge flow toward a junction

Potential Difference a potential difference of 1 volt is produced when one unit of work is done in separating unit charges through unit distance

Power the rate of doing work, or the rate of converting energy. When 1 volt is applied to a load and the current demand is 1 ampere, the rate of energy conversion (power) is 1 watt

Power, Apparent the product of the applied voltage and the current demand of a circuit. Apparent power is measured in volt-amperes

Power, Reactive also called imaginary power or wattless power. Reactive power does no work, but merely surges back and forth in a circuit

Power, Real power developed by circuit resistance or effective resistance

Primary Battery a battery that cannot be recharged after its chemical energy has been depleted

Primary Winding the input winding of a transformer

Pull Box a metal box with a blank cover for insertion into a conduit run, raceway, or metallic tubing that facilitates the drawing of conductors

Pulsating Current a direct current that does not have a steady value

Puncture a disruptive electrical discharge through insulation

Quick-Break a switch or circuit breaker that has a high contact-opening speed

Quick-Make a switch or circuit breaker that has a high contact-closing speed

R symbol for resistance

Raceway a channel for holding wires or cables. It is constructed from metal, wood, or plastics, rigid metal conduit, electrical metal tubing, cast-in-place, underfloor, surface metal, surface wooden types, wireways, busways, and auxiliary gutters

Rack, Cable a device secured to the wall to provide support for a cable raceway

Rating the rating of a device, apparatus, or machine states the limit or limits of its operating characteristics. Ratings are commonly stated in volts, amperes, watts, ohms, degrees, horsepower, etc.

Reactance an opposition to AC current that is based on the reaction of energy storage, either as a magnetic field or as an electric field. No real power is dissipated by a reactance. Reactance is measured in ohms

Reactor an inductor or a capacitor. Reactors serve as current-limiting devices such as in motor starters, for phase-shifting applications as in capacitor start motors, and for power-factor correction in factories or shops

Receptacle also termed "convenience outlet"; a contacting device installed at an outlet for connection externally by means of a plug and flexible cord

Rectifier a device that has a high resistance in one direction and a low resistance in the other direction

Regulation denotes the extent to which the terminal voltage of a battery, generator, or other source decreases under current demand. Commonly expressed as the ratio of the difference of the no-load voltage and the load voltage to the no-load voltage under rated current demand. It is commonly expressed as a percentage

Relay a device that is operated by a change in voltage or current in a circuit, which actuates other devices in the same circuit or in another circuit

Resistance a physical property that opposes current and dissipates real power in the form of heat. Resistance is measured in ohms

Resistor a resistive component; may be of the wire-wound, carbon-composition, thyrite, or other type of design

Rheostat a variable resistive device consisting of a resistance element and a continuously adjustable contact arm

Rosette a porcelain or other enclosure with terminals for connecting a flexible cord and pendant to the permanent wiring

Safety Outlet also termed "ground outlet"; an outlet with a polarized receptacle for equipment grounding

Secondary Battery a battery that can be recharged after its chemical energy is depleted

Secondary Rack a service-wire holder

Seizing Wire a wire that binds other wires together in an electrical connection

Sequence Switch a remotely controlled power-operated switching device

Series Circuit a circuit that provides a complete path for current and has its components connected end-to-end

Service the conductors and equipment for supplying electrical energy from the main or feeder, or from the transformer to the area served

Serving, of Cable a wrapping over the core of a cable before it is leaded, or over the lead if it is armored

Sheath, Cable a protective covering (usually lead) applied to a cable

Shell Core a core for a transformer or reactor consisting of three legs, with the winding located on the center leg

Short Circuit a fault path for current in a circuit that conducts excessive current. If the fault path has appreciable resistance, it is termed a leakage path

Shunt denotes a parallel connection

Sine Wave variation in accordance with simple harmonic motion

Sinusoidal having the form of a sine wave

Sleeve, Splicing also termed "connector"; a metal sleeve (usually copper) that is slipped over and secured to the butted ends of conductors to make a joint that provides good electrical connection

Sleeve Wire a circuit conductor connected to the sleeve of a plug or jack

Sliding Contact an adjustable contact

arranged to slide mechanically over a resistive element, the turns of a reactor, a series of taps, or around the turns of a helix

Snake   a steel wire or flat ribbon with a hook at one end, used to draw wires through conduit, etc.

Socket   an attachment for mechanical support of a device (such as a lamp) and for connection to the electrical supply

Solenoid   a conducting helix with a comparatively small pitch; also applied to coaxial conducting helices

Spark Coil   also termed "ignition coil"; a step-up transformer designed to operate from a DC source via an interrupter that alternately makes and breaks the primary circuit

Sparkover   a disruptive electrical discharge between the electrodes of a gap; generally used with reference to measurement of high voltage values with a gap having specified types and shape of electrodes

Special-Purpose Outlet   an outlet connected to an individual branch circuit, for supplying the current demand of a heavy appliance

Splice   also termed "straight-through joint"; a series connection of a pair of conductors or cables

Split Receptacle   a duplex receptacle in which one outlet is switch-controlled, or a duplex receptacle in which each outlet is connected to a separate branch circuit

Station, Automatic   a generating station or substation that is usually unattended and which performs its intended functions by an automatic sequence

Surge   a transient variation in current and/or voltage at a given point in a circuit

Switch   a device for making, breaking, or rearranging the connections of an electric circuit

Symbol   a graphical representation of a circuit component; also, a letter or letters used to represent a component, electrical property, or circuit characteristic

System   a distinctive set or arrangement of things connected to form a unitary whole

Tap   in a wiring installation, a T joint (Tee joint), Y joint, or multiple joint. Taps are made to resistors, inductors, transformers, etc.

Terminal   the terminating end(s) of an electrical device, source, or circuit, usually supplied with electrical connectors such as terminal screws, binding posts, tip jacks, snap connectors, soldering lugs, etc.

Three-Phase System   an AC system in which three sources energize three conductors, each providing a voltage that is 120° out of phase with the voltage in the adjacent conductor

Tie Feeder   a feeder that is connected at both ends to sources of electrical energy. In an automatic station, a load may be connected between the two sources

Time Delay   a specified period of time from the actuation of a control device to its operation of another device or circuit

Tip, Plug   the contacting member at the end of a plug

Torque   mechanical twisting force

Transfer Box   also termed "pull box"; a box without a distribution panel containing branched or otherwise interconnected circuits

Transformer   a device that operates by electromagnetic induction with a tapped winding, or two or more separate windings, usually on an iron core, for the purpose of stepping voltage or current up or down, for maximum power transfer, for isolation of the primary circuit from the secondary circuit, and in special designs for automatic regulation of voltage or current

Transient   a nonrepetitive or arbitrarily timed electrical surge

Transmission (AC) transfer of electrical energy from a source to a load, or to one or more stations for subsequent distribution

Troughing an open earthenware, wood, or plastic channel in which cables are installed under a protective cover

Tubing, Electrical Metal(lic) a thin-walled steel circular raceway with a corrosion-resistant coating for protection of wires or cables

Tubing, Flexible also termed "loom"; a mechanical protection for electrical conductors; a flame-resistant and moisture-repellent circular tube of fibrous material

Twin Cable a cable consisting of two insulated and stranded conductors arranged in parallel runs and having a common insulating covering

UL abbreviation for Underwriters, Laboratories

Underground Cable a cable designed for installation below the surface of the ground or for installation in an underground duct

Ungrounded System also termed "insulated supply system"; an electrical system that "floats" above ground or one that has only a very high-impedance conducting path to ground

Unidirectional Current a direct current or a pulsating direct current

Units established values of physical properties used in measurement and calculation; for example, the volt unit, the ampere unit, the ampere-turn unit, the ohms unit, etc.

USE abbreviation for underground service entrance

V symbol for voltage, sometimes used instead of E

Value the magnitude of a physical property expressed in terms of a reference unit, such as 117 volts, 60 Hz, 50 ohms, 3 henrys, etc.

VAR denotes volt-amperes reactive; the unit of imaginary power (reactive power)

Variable Component a component that has a continuously controllable value, such as a rheostat, movable-core inductor, etc.

Ventilated a ventilated component is provided with means of air circulation for removal of heat, fumes, vapors, etc.

Vibrator an electromechanical device that changes direct current into pulsating direct current (direct current with an AC component)

Volt the unit of emf; 1 volt produces a current of 1 ampere in a resistance of 1 ohm

Volt-Ampere an electrical unit equal to the product of one volt and one ampere

Voltage the greatest effective potential difference between a specified pair of circuit conductors in a circuit

Voltmeter an instrument for measuring voltage values

Wall Plate see *Face Plate*

Watt the unit of electrical power, equal to the product of 1 volt and 1 ampere in DC values, or in rms AC values

Watt-hour a unit of electrical energy, equal to 1 watt operating for 1 hour

Wattmeter an instrument for measuring electrical power

Wave an electrical undulation, basically of sinusoidal form

Weatherproof a conductor or device designed so that water, wind, or usual vapors will not impair its operation

Wind Bracing a way of bracing to secure the position of electrical conductors or their supports so as to avoid contact caused by wind

Wiper an electrical contact arm

Work the product of force as measured by the distance through which the force acts. Work is numerically equal to energy

Working Voltage also termed "closed-circuit voltage"; the terminal voltage of a source of electricity under a specified current demand; also, the rated voltage of an electrical component such as a capacitor

X   symbol for electrical reactance

X Ray   an electromagnetic radiation with extremely short wavelength, capable of penetrating solid substances; used in industrial plants to check the perfection of device and component fabrication (detection of flaws). For other applications, refer to any standard electrical handbook

Y Joint   a branch joint used to connect a conductor to a main conductor or cable for providing a branched current path

Y Section   also termed "T section"; an arrangement of three resistors, reactors, or impedances that are connected together at one end of each, with their other ends connected to individual circuits

Z   symbol for electrical impedance

Zero-Adjuster   a machine screw provided under the window of a meter for bringing the pointer exactly to the zero mark on the scale

Zero-Voltage Level   a horizontal line drawn through a waveform to indicate where the positive excursion falls to zero value, followed by the negative excursion. In a sine wave, the zero-voltage level is located half-way between the positive peak and the negative peak

INDEX